CASCADES AND FIELDS IN PERCEPTUAL PSYCHOPHYSICS

Previous Texts by the Same Author

Psychometrics of Similarity

Time Series in Psychology

Nonlinear Psychophysical Dynamics

n-Dimensional Nonlinear Psychophysics

CASCADES AND FIELDS IN PERCEPTUAL PSYCHOPHYSICS

Robert A M Gregson
Australian National University

World Scientific
Singapore • New Jersey • London • Hong Kong

Published by

World Scientific Publishing Co. Pte. Ltd.
5 Toh Tuck Link, Singapore 596224
USA office: 27 Warren Street, Suite 401-402, Hackensack, NJ 07601
UK office: 57 Shelton Street, Covent Garden, London WC2H 9HE

British Library Cataloguing-in-Publication Data
A catalogue record for this book is available from the British Library.

CASCADES AND FIELDS IN PERCEPTUAL PSYCHOPHYSICS

ISBN-13 978-981-02-2327-4
ISBN-10 981-02-2327-7

Preface

It is now over ten years since the first work on nonlinear psychophysical dynamics (NPD) was presented; this was partly in Australia and partly at Technische Universität Braunschweig, in Germany. Since then two books, and a number of collaborative papers have teased out extensions and applications of the theory, some in analytical mathematics, most in experimental mathematics (the current name for simulation explorations of analytically intractable processes) and as applications to specific sensory and perceptual phenomena. More recently the extensions into analogies of neural network theory and of cellular automata, and into minimum Boltzmann entropy identification of response surfaces, has changed the emphasis and direction a bit, and approached the questions of process description with more computational diversity.

Parameters in a process model do not have necessarily to match one-to-one in their status or functions any of the parameters in traditional psychophysics. As a consequence one is led to ask different questions, and to meet different conceptual problems in theory construction. One is concerned with retaining some compatibility with what we know about the dynamics of mass action in the brain, though not with the highly specialised peripheral transduction mechanisms of one or two special senses. This concern of course goes back to Fechner's "innere Psychophysik", but the sort of algebra used, putting input-output relations into real time nonlinear deterministic evolutions, and treating inputs as transient destabilisations of an ongoing central equilibrium dynamics, can better find its historical grandfather in Herbart's neglected 1812 paper, on the strength of a sensation as a function of its progression in time.

Using $n\Gamma$ models topics which have been explored include such things as adaptation level in domestic fowls, the size-weight illusion, binary up to quaternary odour mixtures, sensory overload in brain damage, vibrotactile perception on the human forearm, response latency frequency distributions, brightness induction, outliers in the method of limits, neural networks, pattern discrimination, and similarity judgments, and of course coloured pictures of fractals, which exhibit symmetry, have been drawn.

Throughout this study, the pathway between stimulus and response is dual, and the final output is multiple, though only one functionally dominant component may be mapped directly onto observable response properties, at one point in time. It is this fundamental multiplicity of process, at any level from a single sensory channel to a complete connected lattice of simultaneous channels, each dual, which is a main feature distinguishing it from various approaches which have been called *classical* or even *modern* psychophysics. It is not, of course, the only difference, and others are identified or referred back to previous work as they emerge in discussion.

This monograph is the third on NPD, and so far as possible avoids extensive recapitulation of what has gone before, so that some familiarity with previous results and discussion is helpful if not indispensible. This volume focusses on applications, and elucidations of processes, in some very traditional psychological problems which in some cases are of considerable complexity and have a long and detailed experimental history. The Gestalt theories on the 1920s and 1930s came to an end because, though insightful and meticulous in their descriptions of the (mostly visual) phenomenology of perception, there was no way then of extending and testing ideas with computational algebra. What can now be done in 20 seconds of c.p.u. time on a mainframe computer would then have taken, by hand, at least a year and been decidedly error-prone. Modern developments in both computing and nonlinear mathematics were then only on a distant horizon. Now questions and results that may have lain dormant for up to half a century can be revived; the guiding metatheoretical commitment is that the same processes which underlie unidimensional sensory psychophysics could be extended, with minimal ancillary assumptions and reparametrisation, to a few well-explored results in the psychology of perception.

The approach throughout this text has been one of using applied mathematics in a psychophysical context or contexts; the real data drive the applicability of the theory. The reasons for originally considering the potential applicability of the Γ trajectories were reviewed in my two earlier monographs. Over the same period that this work has been done, there have been important developments in areas of pure and applied mathematics that sometimes parallel the explorations of nonlinear dynamics used here. Readers who wish to check this work, or to refresh their acquaintance with formal proofs and existence theorems, should consult the books of Block and Coppel (1991), of Wiggins (1992), and of de Melo and van Strien (1993), and the papers by Branner and Hubbard (1988,1992), all given in the references. I am indebted to Dr Andrew Heathcote (Newcastle, NSW) for referring me to some significant sources on networks, to Professor Hannes and Dr Anna Eisler (Stockholm) for providing more references on time perception, and to colleagues for significant discussions at seminars at A N U. Some of the work described has been reported in preliminary incomplete reviews at conferences in Australia, and in seminars in Europe, including the University of Vilnius, Lithuania, in 1993.

"If we take in our hand any volume.... let us ask, *Does it contain any abstract reasoning concerning quantity or number ?* No. *Does it contain any experimental reasoning concerning matter of fact and existence ?* No. Commit it then to the flames, for it can contain nothing but sophistry and illusion."

David Hume
An Inquiry Concerning Human Understanding, 1739

"mathematics is an instrument which should conform itself strictly to the exigencies of the field in which it is applied."

Bruno de Finetti
Teoria delle Probabilità, 1970

A magnification of a very local region of the Julia set for Γ with $a = 3.615, e = 0.54034$. This is a region of high instability, not used in the simulation of more typical psychophysical processes, during the transition from one to many basins of attraction.

Contents

1 Fundamental Assumptions of Nonlinear Psychophysical Dynamics

"Physicists are not expected to forecast whether the 12.30 from Sydney will be five minutes late, or to explain why a boulder rolling down a slope gets wedged 20 metres from the bottom. Yet psychologists are often accused of being pseudo-scientists because they cannot make analogous prophecies. A scientific law asserts a functional relation betwen variables; it is not a Revelation of Destiny." [after Peters, R. S. (1953) revision of *Brett's History of Psychology, (1912), p 720*].

This monograph is about the application of two mathematical operations, cascades and fields, based on the assumptions of nonlinear psychophysical modelling. In that sense, therefore, it is only concerned with restricted and deterministically-defined forms of cascade and field, but it can be put into a broader historical context which spans many sciences, and relates - albeit loosely - to inbuilt features, and the structure, of mathematical models which are used to simulate and to represent processes in the real worlds of biology and physics.

Because the phenomena of interest are extended in time, never instantaneous, and involve in some cases the simultaneous processing of arrays of sensory inputs in parallel, the term *perceptual psychophysics* has been used, as something of a distinction from the more restricted implied sense of sensory psychophysics which is usual. It might be said with justification that the problems addressed here lie in an intermediate domain between what are traditionally in experimental psychology two areas, each with their own rich and complex archives of data. It is therefore pertinent to make some comments on these apparent distinctions. The distinction between sensation and perception was sharpened into what we might recognize today only in the 19th century, though as Brett (1912) commented "The student of psychology who follows the growth of the subject from Locke to the most recent times will frequently find how utterly these terms cloud the issues". The terms are, of course, sensation and perception; the senses give us knowledge of what is outside ourselves, but we also have inner senses that tell us about events inside our bodies.

The use of *'perception'* as distinct from *'sensation'* is perhaps nicely epitomised by Koffka's question "why do things look as they do ?" Notice that *look* has crept in, giving a modality-specific constraint to the sense of the word 'perception', but then Koffka (1935) was concerned to show that such phenomena as brightness constancy are a breakdown of the simpler sensory psychophysics which start from unidimensional inputs. To quote him (1935, p 90) " there is absolutely no reason why *sensory* equality should *mean* perceptual inequality." He was writing about an experiment which showed that young chicks are subject to Jastrow's illusion. He concluded (page 105) " the true solution.... cannot be a machine theory based on a sum of independent sensory processes, but must be a thoroughly dynamic theory in which the processes organise themselves."

There is a lot buried in that statement which it might be useful to try and disinter. First, a "machine theory" doesn't have to be concerned with a sum of independent sensory processes. The modern neural network theories are capable of describing what happens, or might happen, when you input a collection of independent sensations to a connectionist structure. But what happens there is in no sense a sum, though the neural network is certainly realizable as a machine. Perhaps we should remark parenthetically that Koffka was writing a decade before Turing.

This idea that a collection of inputs is in some way added, like a pack of mixed vegetables sold in a supermarket as 250 gm of carrots, 150 gm celery, 400 gm potatoes, 100 gm broccoli, and so on to make a vegetable soupmix (although the weights are additive the taste sensations in the soup certainly are not), is implicit in the use of the General Linear Model (anova, multiple regression) in experimental designs. No chemical engineer, or even his predecessor a master brewer, would not be in the least impressed by the strange constraint imposed on input effects to yield additivity, but it seems to have taken a great deal of agonising, in the last half century, on the part of psychologists to think past it.

The second point is Koffka's requirement that a theory be *thoroughly dynamic*. A commendable sentiment, but what is dynamic in gestalt descriptions of the end products of perceptual processes ? *Dynamic* should mean that a process goes on in real time, and involves responses to rates of change of inputs as well as to levels of inputs. To be able to do this one needs in turn to have measures on these inputs and outputs, which enable one to say that things are changing faster (or slower) and that the rates of change can be measured independently of their momentary levels.

The third point made by Koffka is that in dynamics, processes "organise themselves". This is a curious notion, because supposedly it means that a process changes the internal transformations it imposes upon a set of inputs as a function of other internal variables, and not as a function of the inputs. But a dynamic system can reorganise itself as a function solely of the momentary inputs, by having more than one sort of internal organisation in its repertoire. Its autonomy is in the way it does reorganizing, but the inputs can trigger the changes in organisation.

However there is some agreement that perception involves inputs that are not unidimensional, and are given some meaning by the observer. Simple sensations are not given meaning except in the sense of the stimulus error; saying 'that is lemon juice' instead of saying 'that is an acidic taste'.

Put bluntly,

Sensations are what we get,

Perceptions are what we make of them.

This folkpsychology definition raises difficulties[1] because of the verbs 'get' and 'make'. 'Get' implies receipt, and transformations intrinsic to receipt and then subsequent transmissions to the higher central processes, 'make' implies more transformations at receipt, and to output. The question is, which transformations are precursors to, and independent of, interpretative transformations after receipt ?

So the question it is necessary to address is, what lies between the traditional psychophysics of simple sensations, and the perception of complex mixture inputs. Is there some other sort of psychophysics which might bridge

[1] The serious problems with this simplistic definition, and the distinctions between psychological and physiological theories of perception, and the phenomenology of form, are elegantly reviewed by Hochberg (1984). The distinctions between what the fields used in this monograph can describe and what psychological transformations are left out of the picture, particularly those concerned with the subsequent or synonymous retrieval of information from a perceived figure, are covered explicitly in his review.

the gap ? Are there sensory inputs which are multivariate, and lead to dynamic reactions (in the true sense of *dynamic*), but without necessarily being given meaning; that is, not being inevitably labelled, interpreted, categorized, or seen as symbolic ?

There is more than a hint about these questions in Gregory's essay "Choosing a Paradigm for Perception" (1974) where he sets up a table and decides the winner is 'perceptions as hypotheses, based on stored and sensed data'. But what is it that is stored, so that hypotheses can be constructed about it, when the inputs are multivariate ? The points that we want to make so far are:

(1) There are gaps in our formal knowledge.

(2) The distinction between sensation and perception is muddy.

(3) In the real world, nearly all sensory encounters give us a string of multivariate inputs, and we respond to them.

(4) The way that the notion of dynamic processes has been used by psychologists in sensation and perception, as opposed to modern physiology, is shallow.

(5) It should be possible to give an account of how mixtures are responded to without any commitment to linear or summatory models, but not otherwise.

(6) If you write nonlinear models you can't use the body of linear statistical techniques to evaluate results, and many statisticians don't anyhow.

It is not novel to say all this; the recent and almost explosive increase in the popularity of ideas from catastrophe theory, neural networks, chaos, and edge-of-chaos, in various disciplines including theoretical psychology, is a response in part to the zeitgeist, and in part to the six points enumerated above.

Using the idea that a three-way distinction, between sensation, apperception (some intermediate stage), and perception, is perhaps a better way to go if we want to distinguish the hardwired from the soft stored operations; ultimately we would like to answer the question, 'are there some common operations which turn up in all sensory modalities when information is integrated before the recipient begins to interpret that information ?' That is, in fact, the question that psychophysics presumes can be answered in the affirmative, by advancing "laws" such those associated with Weber and Fechner, or by Plateau and rediscovered by Stevens. It is not a question that can be answered by considering only unidimensional inputs, but the textbooks' coverage on psychophysics is usually about and only about such inputs. That sensations are not additive was firmly commented on by Herbart (1824, trans. 1891) when he developed mathematical models for sensory interaction, and produced algebras to satisfy the condition that "Of two concepts, no matter how unequal their respective strength, the one can never quite obscure or arrest the other (i.e., drive it out of consciousness); but of three or more concepts, it may happen that one is so weak so as to be entirely arrested by the other two". Herbart is a very interesting thinker because his models are descriptions of processes in real time, not only is the interaction between sensations anything but additive, but it is dynamic and comes and goes. Note that the notion of a limit on channel capacity, which information theorists such as Broadbent in the 20th century made much of, is already there. Let me digress a little more into my favourite 19th century psychologist, and mention his ideas of Wölbung (trans. 'vaulting') and Zuspitzung (trans. 'tapering'). "The effect of a conscious idea on the unconscious idea allied to it is to bring the latter into consciousness in general without accurate discrimination of a congruent part of one which may blend with a part of the other. But with time the incongruent parts are sent back into unconsciousness, and thus the assisted idea is arched, figuratively speaking, in such a manner that its peak value comes into consciousness." It is true that the time course of sensory intensity rises and then falls. To test the Herbartian idea experimentally is not impossible; we could track intensity of one component in a mixture as others with varying degress of similarity are copresented. This curious and unexplored topic (at least in contemporary psychology) is mentioned to show what is possible. It is slightly related to what is explored in tracing processes internal to a psychophysical field, in later chapters.

As a more modern text we may favour von Fieandt and Moustgaard (1977) because they take a European perspective. These two writers are quite firm in damning isomorphism as a Galileo-Newtonian mistake; they note that the doctrine of isomorphism treated stimulus-sensation mappings as one-to-one, and went out with Külpe in the 1890s. They comment "a one-sidedly analytical and elementarist approach to perception is far from being fruitful,and may even lead to a preoccupation with phenomena without relevance for an understanding of the perception of everyday life" (pp. 609-610).

Cross-coupling Channels

Metzger, commented on Helmholtz (1960), who postulated *Urteilstheorie* concerning unconscious inferences. These are unobservable processes which intervene between sensory inputs and percepts, to account for the percepts being other than the linear sums or transforms of the inputs. Metzger (1974) says: "the assumption of unconscious inferences is indispensible only as long as lateral interactions between simultaneous excitations are considered to be impossible". But the assumption of lateral interactions was already made by Hering (1905) and later formulated explicitly by Wertheimer (1912) under the name of *Querfunktionen* (literally 'cross-functions'). Metzger et al (1970) discuss this in more detail.

The qualitative notion of cross-coupling, or mutual asymmetric interaction, which is a particular sort of input-output mapping in n-dimensional nonlinear systems, obviously goes back to the dispute between Helmholtz (1860) and Hering (1905). Helmholtz advanced the notion of intervening processes in what von Fieandt *et al* later called the Galileo-Newtonian tradition. Helmholtz wanted to explain *Urteilstäuschungen* by invoking interpolated "mental activities" that enabled the observer to reconstruct the nature of the external perceptual world from the induced excitations provided by the sensory transducers, these mental activities were inferences that, if they went wrong, produced the *Urteilstäuschungen* as illusions or misperceptions. The genealogy of the idea goes back to Schopenhauer in 1818.

Köhler (1913) pointed out that we don't need (at least in most cases) these unconscious inferences unless we make the mistake of assuming that lateral interactions between simultaneous inputs didn't happen. In short, one sort of assumed linearity forces the theorist into postulating ancillary mechanisms, embodied in an homunculus. The Wertheimer theory is extended by Köffka (1935) in a lengthy and at times impenetrable discussion about traces interacting with process, and exerting mysterious forces on fields. The borrowings from electromagnetism are obvious, and have been used by other perceptual theorists. In visual perception Motokawa (1950) used an electrophysiological field theory to explain some geometrical illusions.

Our objective here is to turn qualitative notions of cross-coupling into algebraic dynamics.

Nonlinear Psychophysics: Fundamentals

The material in this section is taken from Gregson (1988,1992a), where extensive treatment of the rationale and the properties of the models will be found. It constitutes the building blocks on which cascade and field operations have been built; the original definitions of a Γ cascade are in Gregson (1988) and of a field (there called a *network*) in Gregson (1992a).

This section is a slight simplification of relevant theory, the Γ recursion is the core of a family of models, which exhibit a diversity of input-output mappings; in this monograph the starting point is one of the simplest, $\Gamma V7$ in the notation of Gregson (1988). Analytical studies of the Γ recursion and

---

Reference List: Definitions in Nonlinear Psychophysics

1: Γ recursion in a complex cubic polynomial: Single input $\mapsto$ Single Output.

It resembles a classical psychometric function; e.g. weight $\mapsto$ heaviness, and has two parameters; a is a gain, e modifies relative sensitivity to input levels.

Mapping is $U \mapsto a \mapsto \Gamma(Y(Re, Im)|_{e,\eta}) \mapsto Y_{obs}$
where U is a stimulus, and Y_{obs} is a response measure.

(Gregson, 1988)

2: $n\Gamma$ extension to multiple (n) defined inputs; inputs and outputs in one-to-one or many-to-one correspondence.

There is reciprocal asymmetrical interaction between parallel channels in **Case 2**.

For two channels ($n = 2$) the parameters are:

$$a_1, a_2, \qquad e_1 = \lambda/a_2, \quad e_2 = \lambda/a_1$$

For $n > 2$

$$\forall\ x,\ e_x = \lambda/(max(\{a_w\}), w \in W_{(n-1)}, x \notin W_{(n-1)}$$

(Gregson, 1992a)

3: $(n \times n)\Gamma$ multiple inputs n^2 are on a lattice neighbourhood (size $n \times n$); it is $n\Gamma$ algebra but over n^2 channels in a square array.

(Gregson, 1992a)

4: Cascading is an operation of repeatedly taking the output of a Γ or $n\Gamma$ recursion and feeding it back into the a of a new Γ recursion. It is like looping outside and around an inner loop. It can be symbolised as

$$U \mapsto a \mapsto Y_1 \mapsto a_2 \mapsto Y_2 \mapsto a_3 \mapsto Y_3 \mapsto ....a_k \mapsto Y_k \mapsto ...$$

where k is called the *cascade step number*.

The extension of these ideas into the representation of geometrical figures is done by putting the line figures onto some of the pixels of a lattice; two a values, a_0 and a_f are needed initially.. The $(n \times n)\Gamma$ moves over the neighbourhoods in the lattice.

---

its trajectories are in Price and Gregson (1988) and by Campbell and Gregson (1990).

The organism contains an autonomous variable Y which is complex, and runs continuously in discrete time. Inputs destabilise it and it restabilises after an input is finished. There are no random or stochastic processes written into the organism, but the environment can input random or autocorrelated series of arbitrary structure. The time series properties of input stimulus sequences will not in general be replicated in the corresponding output series of responses. The single channel model, Γ, involves the mappings

$$U \mapsto a, \quad \Gamma_{a,e,\eta}(Y(Re, Im)) \mapsto Y(Re)$$

where Γ is the recursion, within one trial, for η steps, of

$$Y_{(j+1)} = -a \cdot (Y_j - 1)(Y_j - ie)(Y_j + ie) \qquad i = \sqrt{-1} \qquad [1.1]$$

for given starting values of Y, and fixed values of a, e. Y is the internal variable of the system, it is never coupled directly to input levels, which only act on the gain parameter a to modify the dynamics of the Γ trajectory. The stimulus input U maps linearly onto a, and the observable output Y(Re) at η maps linearly onto the response. The series of Y(Re,Im) values is called the trajectory of the process, $0 < Y(\text{Re}) < 1$. To avoid degeneracy, chaotic dynamics, or explosions, $2 < a < 5$ is practical. This process should be distinguished from classical psychophysical models in which $Y = f(U) + noise$, with Y scalar and not complex, would be the way of representing stimulus-response mappings in our present notation. As here we are, by definition of perception, in a multichannel situation it requires $n\Gamma$ which has in **Case 2**, as defined next, interaction between channels.

Two cases are defined, which differ in their cross-coupling between channels; the first is simpler and would not generate the anomalies which might, for example, characterise some visual illusions.

Case 1: If the (Im) output of Y_{obs}, (which is the value of Y after η recursions), is taken as interpretable as a form of internal noise, as has been shown to be plausible in other contexts (for example it exhibits intermittency), then let us take the total within-loop noise level at any step in the recursion to be the union of the component noises. So in each cycle the two Y(Im) are replaced by the term $max(Y_{1,j},\ Y_{2,j})$(Im) and as the (Im) components go into limit cycling more readily (that is, at lower system parameter values) than their (Re) counterparts, the effect will depend crucially on the set of values

(a_1, e_1, a_2, e_2) (Campbell and Gregson, 1990). It can be shown, by simulation, that high (e_1, e_2) values induce some curious effects for large U_J (inputs); a paradoxical "turn-over" effect in $U \mapsto Y_{obs}$ graphs for at least one component is observable. At the same time the other component may create a monotonically *decreasing* graph of $U \mapsto Y_{obs}$.

For a double loop put, for **Case 1**, where the two dimensions are denoted by h and i suffices,

$$Y_{hij}(\text{Im}) = \max(Y_{hj}(\text{Im}), Y_{ij}(\text{Im})) \qquad [1.2]$$

and then define

$$Y^*_{hj} = (Y_{hj}(\text{Re}), Y_{hij}(\text{Im})) \qquad [1.3]$$

and

$$Y^*_{ij} = (Y_{ij}(\text{Re}), Y_{hij}(\text{Im})) \qquad [1.4]$$

Case 2: A second cross-linkage is possible if the e terms are made to be functions of the a terms in the opposite dimension. This represents a process of mutual inhibition. Each dimension is desensitised to differences in its own input as the input to the other dimension is increased, viz.:

$$e_1 = \lambda_1 \cdot a_2^{-1}, \qquad e_2 = \lambda_2 \cdot a_1^{-1}, \qquad [1.5]$$

where $\lambda_1 > 0$, $\lambda_2 > 0$, and $2.6 < a_{min} < 4.4$ is a working range for $0.05 < e < 0.40$, and approximately $.20 < \lambda < 1.2$ will avoid the explosive condition (Gregson, 1988, Chapter 2) of $ae > 1.7$.

The advantage of writing [1.5] is that the number of parameters in the the model is reduced by one, e_1 and e_2 are removed and λ is introduced, with $\lambda_1 = \lambda_2 = \lambda$. The assumptions [1.2 - 1.4] are in already in **Case 1**, and for **Case 2**, additionally we now have

$$Y_{h(j+1)} = -a_h \cdot (Y^*_{hj} - 1)(Y^*_{hj} - i\lambda_h \cdot (a_i)^{-1})(Y^*_{hj} + i\lambda_h \cdot (a_i)^{-1}) \qquad [1.6]$$

and correspondingly

$$Y_{i(j+1)} = -a_i \cdot (Y^*_{ij} - 1)(Y^*_{ij} - i\lambda_i \cdot (a_h)^{-1})(Y^*_{ij} + i\lambda_i \cdot (a_h)^{-1}) \qquad [1.7]$$

The 2Γ models extend to higher dimensionality (Gregson, 1992a) and could be applied, for example, to triple sensory inputs.

The Γ V 10 model, which in Gregson (1988) is set up only for two channels, uses [1.1] for each channel to give $Y_1(\text{Re})$ and $Y_2(\text{Re})$ at the end of the first trajectories. Then the output Y_c from the second trajectory in the cascade is given by

$$Y_c = \Gamma(Y_1(\mathrm{Re}) + Y_2(\mathrm{Re}) - Y_{1,2}(\mathrm{Re})) \qquad [1.8]$$

where

$$Y_{1,2}(\mathrm{Re}) = max(Yb_1, Yb_2), \qquad [1.9]$$

$$Yb_i = \Gamma(a_i^*, e, \eta), \quad i = 1, 2 \qquad [1.10]$$

and

$$a_i^* = min(a_i) + \delta a, \quad \delta \sim N(0, \sigma_{ai}) \qquad [1.11]$$

where σ_{ai} represents variation (in time), of a source U_i and not intrinsic variation within the Γ parameters.

There is no cross-coupling in Γ V 10, the cascade operation is an alternative to the role of λ in 2Γ **Case 2**. The 2Γ models have no cascading.

Equivalent Representations in Nonlinear Psychophysics utilising Volterra-like Kernels for Gamma Recursions

There are other ways of thinking about the processes assumed here, which tie in with ideas in statistics or in artificial intelligence; they are mentioned in this and the next section to put the approach in context for readers who start from outside psychophysics. The Γ recursion is what physicists sometimes term an *evolution equation*. Evolutions are ubiquitous in, for example, theories of wave propagation in hydrodynamics, and have been extended to biophysics (Engelbrecht, Fridman, and Pelinovski, 1988). Such equations are often analytically intractable, and are treated by piecewise linearisation, or by invoking the centre manifold theorem, starting with partial differential equations (PDEs), as in Altman (1986). Even if generalised approaches are proposed, the mathematical conditions for obtaining existence theorems, or solutions to such properties as first passage times, appear to be restrictive and we have no sure advance information concerning what conditions are reasonably to be met in psychophysical models. It is now well known that the sort of problems we are faced with in psychophysical theory are in the mathematical sense ill-posed (Keyfitz and Shearer, 1990); the most we can be sure of is that the psychophysical evolution from stimulus to response has a dissipative component and is nonstationary. It is possible that we might make some headway by invoking Hopf's (1954) theorem, as the Γ recursion can be coarse-grained as a

Markov chain (Gregson, 1993a) which is always nonnegative in its parameters, $a > 0, e > 0$ (Nummelin, 1984)[2].

The Hopf theorem shows that the state space of a Markov chain on a finite measure can be divided into two parts, the first of which is dissipative. The second part can take one of three cases; if it is null the chain is dissipative, if it is less than zero the chain is absorbing, and if it is greater than zero then it is absorbing and does not contain any positive dissipative subset.

As a Γ recursive chain changes its form at the bifurcations, we might hypothesise that its Markov representation (which includes a kernel in the form of the transition matrix) changes as the system bifurcates. One way to consider a kernel representation of Γ recursions is to compute an expression for the autocorrelation spectrum over a local region, and then from that obtain the transfer function coefficients and hence their polynomial factorization. This, however, raises the difficulty that we will need the analog of a correlation coefficient in complex variables. The statistical treatment of complex variables goes back at least to Wooding (1956). Miller (1968,1974) has proved the equivalence of the moments of a complex gaussian process and a bivariate gaussian process, and Kabe (1966) found analogues for other distributions. So we could use the bivariate representation of the complex trajectory and then compute a multiple R in two dimensions[3].

A useful result upon which we may draw is by Miller and Rochwarger (1970). We begin by noting some properties of the lag-one autocovariance of the Γ recursion: Multiply both sides of [1.1] by $(1 + Y_j)$, giving

$$Y_{j+1} + Y_{j+1}Y_j = a(1 - Y_j^2)(Y_j - ie)(Y_j + ie) \qquad [1.12]$$

and expand the complex product $Y_{j+1}Y_j$ within the covariance as

$$cov(Y_{j,j+1}) = (Y_j(Re)Y_{j+1}(Re) + Y_j(Re)Y_{j+1}(Im)$$

$$+Y_j(Im)Y_{j+1}(Re) + Y_j(Im)Y_{j+1}(Im)) \qquad [1.13]$$

---

[2] The approach of Nummelin does support extension into irreducible kernels, which can be non-Markovian (See Nummelin, 1984, p 141).

[3] Actually there exists an extensive literature on the statistics of complex variates, because they are apparently useful in electrical engineering to represent transmission in noisy circuits. See McGee in *IEEE Transactions on Information Theory IT-17*, 1971, pp 149-157.

As before the first bifurcation the recursion runs to an attractor on the real line, the autocorrelation runs in the limit to unity, the terms in (Im) vanish for large η. It is not this limit which is of interest, but the dynamics before stability.

Consider the forward (j to $j+k$) autocorrelation (ar) at lag k over a recent time segment length s, utilising the product term in [1.12], using relaxation over the neighbourhood $\delta a, \delta e$ of the system parameter space in a, e,

$$ar(Y_{j,j+k}|\delta a, \delta e, s) = \frac{cov(Y_{j,j+k})}{sd(Y_j) \cdot sd(Y_{j+k})} \qquad [1.14]$$

This $ar(Y_{j,j+k})$ for a series of s-length windows is itself a time series over j; for each window location we have an one-sided autocorrelation spectrum $\mathcal{S}|s,j$. It is the stationarity of that $\mathcal{S}$ which is of interest; if it is markedly nonstationary and cannot be made stationary by differencing, time series approaches are ruled out, though the nonstationarity itself is interesting. This seems cumbersome, and does not attack the problem with sufficient directness.

Instead, suppose that [1.1] is taken as the kernel of the generation of lag-one transitions and, using operator notation, expanded to get as far as is needed the kernel series

$$\Gamma^1(Y_j),\ \Gamma^2(Y_j) = \Gamma^1(\Gamma^1(Y_j)),\ \text{and recursively to } \Gamma^\kappa(Y_j) \qquad [1.15]$$

where κ is the order of truncation of the kernel series. Obviously, substituting in [1.1] with $j+1$ for j, etc.,

$$Y_{j+2} = a(1 - Y_{j+1})(Y_{j+1} - ie)(Y_{j+1} + ie) \qquad [1.16]$$

and $\Gamma^2(Y_j)$ is

$$Y_{j+2} = a(1 - a(1 - Y_j)(Y_j - ie)(Y_j + ie))(a(1 - Y_j)(Y_j - ie)(Y_j + ie)$$

$$-ie)(a(1 - Y_j)(Y_j - ie)(Y_j + ie) + ie) \qquad [1.17]$$

which involves terms $\mathcal{O}(Y_j^6)$. As $0 < Y < 1$ by definition, some of these higher orders of Y_j are going to be negligible, if the powers of a, ($a > 2$) increase more slowly than those of Y, and $\kappa > 3$ can be dropped in a noisy system, though in a deterministic chaotic part of the phase space it obviously is not so.

Table 1.1; coefficients of recursive Γ operators.

$a^{-2}\Gamma^2$	coefft.	$a^{-3}\Gamma^3$	coefft.
Y_j^0:	e^4	Y_j^0:	e^6
Y_j^1:	$-2e^4$	Y_j^1:	$-3e^6$
Y_j^2:	$2e^2+e^4$	Y_j^2:	$3e^4+3e^6$
Y_j^3:	$-4e^2$	Y_j^3:	$-9e^4-e^6$
Y_j^4:	$1+2e^2$	Y_j^4:	$3e^2+9e^4$
Y_j^5:	-2	Y_j^5:	$-9e^2-3e^4$
Y_j^6:	1	Y_j^6:	$1+9e^2$
		Y_j^7:	$-3-3e^2$
		Y_j^8:	3
		Y_j^9:	-1

Collecting coefficients of powers in [1.17] for Γ^2, and analogously for Γ^3, we have Table 1.1.

The coefficients in Table 1.1 could be used to compute longer lag autocovariances as in [1.13,1.14], if linearity at the level of $\mathcal{F}$ is accepted, where $\mathcal{F}$ is the filter representation of the system. From Table 1.1, given the range of Y_0 and e values usually substituted in [1.1], the expansion of Γ^3 would be restricted to non-negligible terms in the region of Y_j^4 through Y_j^6, and for Γ^2 to Y_j^2 through Y_j^6. As a filter we can put (dropping the j suffices for Y), the superposition of component filters as

$$\begin{aligned}\mathcal{F} =&\Gamma^1+\Gamma^2+\Gamma^3\\ =&f(Y^0,Y^1,Y^2,Y^3)+f(Y^2,Y^3,Y^4,Y^5,Y^6)+f(Y^4,Y^5,Y^6)\end{aligned} \qquad [1.18]$$

What could this represent?

The Γ model is for a single input, at a point in time (a Dirac delta function in some terminology), triggering an evolutionary trajectory that after Y_0 is autonomous from the environment that creates the input onto a. That is, inputs are not sustained in time. Overlapping or temporally extended inputs are considered from one computationally simple viewpoint in Gregson (1992a), but here a sustained input superposing trajectories is observed: $Y_J, (J=1,2,3,\delta J=mj)$ with m unknown a *priori*. This resembles a Volterra kernel (Schetzen, 1980). However, the procedure here is the reverse of that involved in kernel identification, given only input-output mappings from impulse responses (Kailath, 1981). Here the theoretical kernel is known already, so the problem is to see if it can match observed psychophysical data or if it needs extra convolutional terms. In psychophysics there is not usually an analogue of system control through filtering, but to achieve optimal control (via feedback) a knowledge of [1.18] would be needed.

As noted in [1.18], the extended Γ response at one point is derived by superposing individual (nonlinear) trajectories lagged relative to the origin at $j = 0$. The question is whether in fact this superposing, which implies linearity at the $\mathcal{F}$ level, is itself correct. If it is wrong, then the decomposition of an output into the series $\mathcal{F}$ will not hold, and instead it is necessary to rewrite $\mathcal{F}$ as

$$\mathcal{F} = \Gamma^1+\Gamma^2+\Gamma^3+\theta_{12}(\Gamma^1,\Gamma^2)+\theta_{23}(\Gamma^2,\Gamma^3)+\theta_{13}(\Gamma^1,\Gamma^3)+\theta_{123}(\Gamma^1,\Gamma^2,\Gamma^3) \quad [1.19]$$

and following Marmarelis and Marmarelis (1978, p 138) these θs denote deviation from superposition, necessitating therefore a higher-order kernel for process representation.

The form [1.19] suggests a type of experiment in which an independent variable is input duration. If the θ are nonzero then lengthening the time span of the input stimulus induces nonlinearities at the level of $\mathcal{F}$, and [1.18] would fit progressively worse as m increases.

Networks for Nonlinear Multiple Regression compared with n-Gamma for One Output

One particular class of neural networks, termed Parallel Distributed Processing (PDP) in cognitive psychology, has been extensively studied as models of memory and decision making. There are analogies and differences between such networks and those used here, so the distinctions have to be drawn.

If a network with minimally three layers, and with linear elements at the input and output layers, and ogival elements in the hidden layer, is used to map n inputs onto a single output, it can reproduce a wide diversity of forms of nonlinear regression (Mielniczuk and Tyrcha, 1993). No constraints on the linkages between nodes are initially imposed and the network is fully connected. The general functional form being modelled is

$$y = f_n(x_1, x_2, x_3, ..., x_n) \quad [1.20]$$

where f_n is of indeterminate order, and bearing in mind the forms of the elements in the three layers, [1.20] can be expanded to

$$\forall\ x_i,\ x_i = \alpha_i + \beta_i \cdot u_i \qquad [1.21a]$$

where α and β are scalars,

$$v_k = \Phi(x_i, ...) \qquad [1.21b]$$

for the middle layer where Φ induces ogival-like mappings[4], asymptotic to zero and one at the limits and hence non-linear, and

$$w = f(v_k, ...) \qquad [1.21c]$$

where f is linear analogously to [1.21a]. The number of nodes (the *cardinality*) in the middle layer set can be very high if a capacity to reproduce regressions to arbitrary accuracy is demanded of the network. Computationally this can induce very slow convergence in the training phase of the network. The single final output is a linear summation

$$y = \sum c_h w_h \qquad [1.21d]$$

where the c_h are scalar multipliers. The only externally-observable terms in this model are the set of inputs$\{u_i\}$ and the single output y.

The mappings are

$$u \mapsto x \mapsto v \mapsto w \mapsto y \qquad [1.22]$$

because in this state, after equilibrium of throughput has been attained, there is only feedforward. Even if only three layers are used, universal approximation to an unknown mapping and its derivatives is possible under mildly restrictive conditions (Hornic, Stinchcombe and White, 1989,1990). It is worth commenting that human psychophysical judgments cannot attain arbitrarily-defined accuracy.

The question examined here, is to what extent an $n\Gamma$ model with one output Y_{TOT} which is the sum of n internal Y outputs $y_i, i = 1, .., n$ resembles in structure and function the network of [1.22] ?

---

[4] Alternative middle-layer substitutions of different functions at nodes are reviewed by Ripley (1994); an important difference here is that the Y function is complex, but in PDP networks functions in the reals are usual.

The $n\Gamma$ case with one output is used to map the total perceived intensity of mixture stimuli, where the mixture is known *a priori* to have exactly n physical components. Clearly it can also be used to predict component intensities from $\forall\, h:\ w_h \mapsto Y_h$ or relative intensities ($relY$) from

$$\forall\, h:\ w_h / \sum w_h \ \mapsto relY_h \qquad [1.23]$$

As the mapping of inputs U_i onto a_i gains is taken to be affine, this first stage in $n\Gamma$ does parallel, in a constrained way, [1.21a]. The number of the Γ recursions is obviously n, so the number of nodes at the second nonlinear stage is the same as the number of input nodes. This is an important difference from the neural networks general case. The network is connected within the nonlinear stage via the cross-couplings λ in [1.5] and all $y_h (\simeq y_i)$ converge on the final sum. The connectivity of the lattice here is complete only in the sense that each of the n channels is potentially dominated by its largest competitor in the remaining $n-1$ channels, but the form of coupling is defined in a restricted manner and involves complex variables at each node. It is this $n\Gamma$ **Case 2** structure that comes most closely in its functions to the middle layer operations of PDP neural networks. Neural network connectivity is not usually defined in terms of complex variables at nodes or on links. This final convergence is again linear and thus parallels [1.21d]. However, the degree of connection within the $n\Gamma$ mapping is predetermined both in terms of what links exist, and what values are attached to them, and the parameters of the nonlinear transformations are not tuned in the sense of a network which learns. There is no feedback in an $n\Gamma$ (or an $(n\times n)\Gamma$) network, other than the intrinsic recursion of the Γ trajectories, unless the network is cascaded, but such cascading is not necessarily of itself matched to an externally-defined target configuration. This is distinct from cases induced by backpropagation. In other words, as has been previously emphasised, the two models, though having a structural resemblance, are designed to do a different job from each other. The question immediately raises itself, if we leave the $n\Gamma$ parameters as $\{a\}, \{e\}$ (this can arise either by having no cross-coupling as in **Case 1**, or by having all cross-couplings λ in [1.5] independently parametrised) which now raises the number of model parameters in the nonlinear middle mappings from $n+1$ to $2n$, does this relaxation of constraints make it possible for the $n\Gamma$ structure to begin to function like a system capable of learning ? In other words, if we use neural network programs with adaptability to act as a context for an $n\Gamma$ system with *a priori* undefined parameter values, but have some observable $U \mapsto Y$ data for an objective, will this neural network converge on a parameter configuration which is in some sense optimal, and will this configuration itself resemble the $n\Gamma$ forms we have previously used ? Will in effect the parameter dimensionality, associated with the cardinality m of the hidden layer, shrink from the initial $m >> 2n$ and converge onto $n+1$?

Adaptability versus Dynamic Stability

The analogies and contrasts so far noted are focussed on structural considerations. There is however another basis for comparing the two approaches, because neural networks are designed to learn, to categorise, and to remember distinctions between inputs, whereas $n\Gamma$ are not given any learning capacity. The input-output mappings onto which NPD converge are invariant, given the parameter values, because Γ trajectories are deterministic though dissipative and irreversible. This lack of learning capacity is deliberate, and intended to reflect the fact that sensory learning in terms of shifting thresholds, or altering the shape of psychometric functions, is very limited. In contradistinction, learning to give names to sensations is potentially provided for in GZE (Gregson, 1993a) where a neural network, downstream from and taking $n\Gamma$ outputs as its own inputs is used. One may thus think of the GZE structure as a compound of $n\Gamma$ and a network whose cardinality in its hidden layer is not necessarily related one-to-one to n. The functional equivalence of the GZE structure, in terms of learning capacity, to a three-layer neural network with indeterminate and weakly constrained hidden-layer cardinality suggests that the formal proofs of neural network capacity to approximate any regression are supportive of the idea that GZE would also exhibit flexibility in the learning or classification of sensory qualitative inputs, and even linkages between quality and stimulus input levels, but not the unconstrained generalisation of the networks with open hidden-layer structure. The latter are probably too rich to be plausible psychophysical models.

Synopsis

This chapter has sketched the historical context of the problems addressed, and restated the core of the dynamics to be used throughout the remaining chapters. Analogies with, and differences from, other nonlinear models sometimes used in psychology have been indicated.

The next two chapters are concerned with some statistical methods of pattern matching, and with extended properties of Γ trajectories and cascades, that are used later in various cases related to perceptual analyses. These have not been covered in previous NPD work.

Chapter 4 reviews the properties of some unidimensional cascades employed in modelling the psychophysics of time interval estimation; cascades here involve the restabilisation operations already defined in the previous chapters.

In chapters 5 and 6 fields, extended over bivariate Γ lattices, and also using cascading operations, are explored in detail. First the changes in generated contour maps, using spline fitting to networks, as the $n\Gamma$ process moves in its phase space towards chaotic dynamics, are mapped. The the processes are examined again in detail in a series of special configurations that occur in the well-known visual illusions. It is here that contrasts with Gestalt theory are emphasised.

Chapter 7 reviews more applications with slightly different structures. Similarity contour maps are treated as a derived field, extending the analyses of chapter 3 on separability, without invoking metric space geometries. Abrupt discontinuities in switching between series, yet another type of cascade, are examined, and the problem of induction between adjacent fields is considered.

The final chapter 8 reverses the perspectives of the study, by instead addressing the question of whether there is any evidence about dynamics in brain neurophysiology which could support, or at least resemble, the assumptions of NPD theory. Psychophysical theory is terse algebra for observable invariances in the mappings betwen physically-definable external stimulus properties and human experiential responses, but the more detailed the psychophysics becomes, the more dynamical evolutions in time are postulated between stimulus and response, the more it is necesary to ask about the plausibility of the mathematics. There is no sense, nor justification from considerations of tractability, for inventing operations on inputs which we know the brain does not do, or could not do. A psychophysics which ignores the dynamics of its neural substrate is only artificial intelligence, where the theorist is given an excessively free hand to speculate. The interest here is to see if, at the level of system dynamics, a bridge can be built between levels of description, without espousing a rigid reductionism. A schematic diagram, summarising the different sorts of theory and information that direct the tasks of theory construction and modelling in psychophysics, is appended as a sort of framework or map of where the arguments seem inevitably to lead.

Relationships and Analogies

Parallel Distributed Processing ↘ | | Cellular Automata Diffusion ↙

nonlinear multiple regression | Observable Psychometric Functions | figure-ground noise diffusion

↖ ⇑ ↗

Nonlinear Psychophysical Dynamics

↙ ⇓ ↘

phase space and attractor representations | ⇓ ⇓ ⇓ | multiple transient hierarchical recursions

↗ ⇓ ↖

Global Topology of Cubic Polynomials | Γ
 $n\Gamma$
 $(n \times n)\Gamma$
 Trajectories | Cortical Sensory Neuro-physiolology

In this diagram, Γ is for a single channel, $n\Gamma$ is for a set of channels in parallel, and $(n \times n)\Gamma$ is for lattices which are spatially extended in two dimensions. Any of these Γ trajectories can be cascaded, but only the lattices support fields.

2 Nonlinear Psychophysics, Response Surface Identification, and Cross-Entropy

In this chapter some methods for comparing theory with data in later chapters are set out, and a case study in tactile psychophysics is described. In simple experiments, where a set of predetermined stimulus values is used in random order to generate one set of (average) response values, the results are sufficiently expressed on a graph on two axes. In nonlinear psychophysics the independent variables are at least two; a as gain and e or η as internal parameters. The plot then has to be a surface of $Y(\mathrm{Re})$ altitudes, with a, e or a, η as base plane axes. Real data are then compared with one or more sections through the surface. As soon as the dimensionality goes up, to $n\Gamma$ systems, 3-d section surfaces through a hypersurface are the most that can be visualised. All these are called, collectively, theoretical response surfaces. Empirical nonlinear psychophysical systems exhibit complex dynamics, so that their response surfaces can have intricate shapes, and are consequently readily misinterpeted as being heavily contaminated with stochastic noise. The shapes, or better the topology, of these surfaces can be variable and analytically intractable, if it desired to avoid making strong simplifying assumptions.

In the Γ theory of nonlinear psychophysical dynamics, where the generating axes are relatively simple functions of the system parameters, the same holds, but the response surfaces are strictly reproduceable. In both data and in theory parts of these surfaces may be featureless and uninformative, whilst other parts may be very convoluted with steep and locally rapidly changing gradients. It follows that theoretical predictions of input-output mappings are not readily utilisable in matching data to theory, unless we have at least an approximate notion of where on the response surface the observed data are supposed to lie, and have an algorithm for determining relative goodness of fit, of the empirical to the theoretical response surfaces, in the appropriate locations implied simultaneously by both data and theory.

This problem is approached by drawing on two different conceptual frameworks, none of which are in general use in experimental psychology, but all of which now have an extensive literature. These are maximum entropy (ME) modelling, and Bayesian statistics. The first in fact goes back to the nineteenth century and the work of Boltzmann and Gibbs in statistical thermodynamics (Jaynes, 1957), the modern developments in Bayesian inference are due to Jeffreys, Lindley, and Savage (Skilling, 1989a), and nonlinear psychophysics is more recent than either. Some very tentative anticipations of the current approach appear in statistical texts by Schmitt (1969) and McGee (1971), sources directed at social scientists. Interestingly, Dale (1991) identifies maximum entropy as an extension of inference axioms used by Donkin (1851).

Here we are working towards the analysis of very noisy psychophysical data, for which a precedent exists in the analysis of nearly black objects in image reconstruction (Donoho, Johnstone, Hoch, and Stern, 1992). Such problems are said to *ill-posed*, in that in contrast with regression theory, the effective dimensions of the data are much fewer than in the theory, so that the theory predicts a few singular values significantly different from zero. Psychophysical data have also, or can be transformed to have, nonnegative outputs over all their input spaces. De Groot (1970) had observed (p. 431, section 14.16: uncertainty functions and statistical decision problems) that the entropy function is an uncertainty function for nonnegative concave functions defined on the set of all possible distributions of a parameter W generating the probability densities.

In the discussion of a paper by Aitkin (1991, pp. 128-133) contributions by Barnard, Lindley, Cox, R. L. Smith and H.M. Smith drew out distinctions between

(a) the relative fit of two models, assuming that one is "true", and

(b) analyses of the adequacy of fit of M_1, looking for departures in the direction of the other M_2, and vice versa. Here it does not follow that only one fits, both or neither may fit data.

It is the second case upon which we focus here, because quite specific models can be created, but only given a prior probability ($\neq 0, \neq 1$) of the applicability of a model. In this sense we are strictly Bayesian, as is ME.

The algorithms for ME optimisation are iterative, and the functions minimised (or maximised) are often of the form of penalised least-squares, involving the estimation of a regularisation constant K (Davies and Anderssen, 1986). In general, where K is a unit conversion constant, p_i is the probability of the

i^{th} event, H/K is maximised in the information expression

$$H = -\sum_{i=1}^{h} p_i log_e(p_i) \qquad [2.1]$$

subject to the constraints that

$$\sum_{i=1}^{h} p_i = 1, \qquad K = \sum_{i=1}^{h}\sum_{j=1}^{m}(p_i g_{ji} = K_j) \qquad [2.2a, 2.2b]$$

where K_j, g_{ji} are constants; m is at least two less than the number of possible outcomes of p.

When a model M exists to generate a full prior distribution over the system parameter space, then for any variant of M specific predictions m_i (themselves like probabilities) exist for each point of the h in the 1-d response (ψY) space in which the data lie. In this case the minimisation of the cross-entropy is sought, which is defined as

$$B = -\sum_{i=1}^{h} p_i log_e(p_i/m_i) \qquad [2.3]$$

where now i is the counting variable on the points. It is worth noting that the order of enumeration of the pixels is uninformative, the correspondence between p and m is what matters. The cross-entropy B is sometimes known as Boltzmann entropy, and in information theory as the Kullback-Leibler measure.

The m_i are everywhere definable in the parameter space over which the summation in [2.3] ranges, but the surface[1]. defined by $m_i \in M$ is not necessarily everywhere continuous in its derivatives, it can have, for example, boundaries, cliff edges or cusps, or multiple representations at a point where the inbuilt dynamics have gone past one or more bifurcations.

The p_i are data only defined where they are observed, in general on a small fraction of the coordinates of the response surface corresponding to the M surface To employ B an implicit assumption, that the continuity of the response

[1] Here M is used to name a model, and $\mathcal{M}$ is used to name the set of m^2 points on which M can be defined. In experimental design terms, the p_i may be at fixed levels and cells of a full factorial layout. At any point they can have a frequency distribution which is not necessarily unimodal centrally tending (for example, see Gregson, 1992b). In linear psychophysical theory the m_i are single valued, and some average p_i is used to match response surfaces.

surface would be at least the same as that of M, if M existed, has to be considered. If from Y_{obs} a complete surface is constructed (as in later chapters on fields) the local continuity is created by using bivariate spline fitting which does its best to make the first and second derivatives continuous. This is a stronger constraint than need occur if M is generated by processes such as $(n \times n)\Gamma^k$, (see chapters 5 and 6), which may resemble in their dynamics the unidentifiable products of cellular automata. In traditional psychophysics it would be usual to postulate a continuous if bounded M surface and treat discontinuities in responses as the consequence of breakdown (such as the failure of Weber's Law), or just error[2]. The axioms[3] of consistency (Shore and Johnson, 1980) which constrain the solutions are as follows:

Axiom 1: Uniqueness. There shall be only one solution which, up to trivial transformations, satisfies the requirement of being representational.

Jaynes (1957) observes that the ME distribution is uniquely determined as *the one which is maximally noncommittal with regard to missing information*, and agrees with what is known.

Axiom 2: Invariance. Coordinate systems in data or theory do not matter.

Axiom 3: System independence. It should not matter whether one accounts for independent data sets from independent systems separately or in terms of a joint probability density.

Axiom 4: Subset independence. A subset of system states can be treated equivalently in terms of a separate density or in terms of a full system density.

These axioms have an intuitive correspondence with ideas which are implicit in experimental design in psychology. When a model, which is a mathematically formalised expression of a general theory in psychophysics, is fitted, it is assumed that the model structure, though not its precise parameter values, is invariant over subjects and over sessions. This holds even if we postulate *a priori* that learning, transposition or adaptation effects prevail over time, because such processes and their consequences are intrinsic to a well-formed

---

[2] It should be noted that B is intrinsically asymmetrical, so it does not define a distance between two surfaces; distance measures, effectively representing the volume trapped between two surfaces, can be defined.

[3] This writer would prefer to call them *boundary conditions at the metatheoretical level*, they are not like axioms in a mathematical proof as they are given here.

model and inbuilt as slower parameter drifts or shifts as over time.

The single feasible model response surface (Y_{obs}) over the set of Γ psychophysical models is the M_Γ which has the greatest configuration entropy. From maximum entropy theory it is known, as Gull and Skilling (1984) state, that " From axioms of transformation invariance and system and subset independence ... the only consistent variational principles are those which maximise functionals of the form $S = -\sum_{i=1}^{h} p_i log_e(p_i/m_i)$ or (trivially) monotonic functions of S."

There is no need for the p_i or the m_i to be probabilities, they can be any set of proportions. "The application of entropy to the pattern of proportions of intensity in an image seems very natural." (Skilling, 1984). We are concerned with the configurational properties of the response surface, and not its absolute values in ψY units for real psychological data, or corresponding Y_{obs} units from $n\Gamma$ theory.

As a point of interpretation in posterior surface identification, it is critical to note that the maximum entropy image is not more likely than any other feasible image, but with the available information we can make no other consistent choice. What is obtained from the ME analysis is not a variant of $n\Gamma$, unless the analysis is so constrained, but a distribution of maximum entropy which is a compromise between observations and the best-fit theory in the ME sense.

This idea of model consistency corresponds to **Axioms 3** and **4** . If a theory $M_{n\Gamma}$ exists, it will make predictions m_i over all the pixels in an output space

$$Y_i = f(\Gamma\{a\}_{g=1}^{n}; \lambda, \eta) \qquad [2.4]$$

for a defined parameter space in λ, η, because the pixels correspond to the set of chosen parameter values which are the discrete values used in computer simulation. The dimensions $g = 1, 2, ..., n$ of the Γ model are the dimensions of the process generating the observed Y_{obs} surface; they are not the fractal dimensionality D_2 of the output surface in Y. We may need to compute D_2 to identify local dynamics in the parameter space, as an aid to setting up preliminary correspondences between the areal span of the data response surface and the span of the model parameter space which is most relevant to observations.

In data collection, however, an exhaustive mapping of responses over the independent stimulus variables U which are selected as affine transformations of $\{a\}$ is not usually feasible. Fortunately, in maximum entropy modelling it is not necessary either (Donoho *et al*, 1992). The missing cells in the data pixel matrix, which are those where theory exists and data do not, are replaceable with maximum entropy values under the boundary constraints operating. These can be allocated randomly, with low variance, and will have an exponential distribution (one of the results from classical statistical thermodynamics). To formalise, sparse data sets $\{x\}$ (e.g. as observed in Donoho *et al*, 1992) are sparse if and only if some values exist for which

$$\begin{aligned} x_1 &\ correspond\ to\ a_1\ in\ \Gamma \\ x_2 &\ correspond\ to\ a_2\ in\ \Gamma \\ d_{xx'} &\ correspond\ to\ \lambda\ in\ \Gamma \end{aligned}$$

where $d_{xx'}$ relates to a spatial or functional separation of two inputs x, x'. By iterative exploration of preidentification of the stimulus parameter ranges which apparently maximise

$$\frac{\partial \psi Y}{\partial x} \qquad \frac{\partial \psi Y}{\partial d_{xx'}} \tag{2.5}$$

we focus on those psychophysical regions where response differentiation to stimulus change is greatest; in other words on regions of maximal sensory sensitivity[4]. The mapping of $n\Gamma$ onto the data set is thus constrained into those regions as the remainder are a *priori* uninformative both from the viewpoint of theory and of known psychophysical data sets; behaviour which is consistently in sub- or supra-threshold regions is not useful except as an estimate of the variance of quasi-random or purely response-determined behaviour.

Entropy surfaces have not been computed in any of the previous work on $n\Gamma$ processes, though information theory measures on Γ were cited as one diagnostic procedure for identification in Gregson (1988).

---

[4] If a response surface is differentiated, a sensitivity surface is obtained. This has peaks where rates of change in Y_{obs} are maximal, and thus shows regions of greatest response (i.e. information) uncertainty. In a $\Gamma_{(a,e)}$ psychophysical escarpment surface the corresponding sensitivity surface has a ridge along the e values.

Algorithm Structure

The maximum entropy estimation requires iterative computation for its solution. In this section some aspects of the treatments by Delsuc (1989) and by Mohammed-Djafari and Demoment (1989) are followed.

The entropy of the model's image is

$$H_k = -\sum p(Y_{\Gamma,k} \cdot log(p(Y_{\Gamma,k})) \qquad [2.6]$$

and

$$H_{n\Gamma} = \sum_{k=1}^{m^2} H_k \qquad [2.7]$$

It is assumed that the experiment can be performed in a number of exchangeable replications.

a: If only average response values γ_k are available and all can be taken as positive, the maximum entropy distribution of the $p(Y_{\Gamma,k})$ is

$$p(Y_k) = (1/\gamma_k) \cdot exp[-(Y_k)/\gamma_k)] \qquad [2.8]$$

and the maximum entropy H_k is

$$max H_k = 1 + log(\gamma_k) \qquad [2.9]$$

so

$$max H_k = \sum_{k=1}^{m^2}(1 + log(\gamma_k)) = m^2 + \sum_{k=1}^{m^2} log(\gamma_k) \qquad [2.10]$$

b: If there exists some prior knowledge on the noise, and it can be assumed that the density function of $Y_{n\Gamma}$ is exponential,

$$H(Y_{n\Gamma}) = -\sum_{k=1}^{m^2} q_k \cdot log(q_k), \qquad q_k = Y_{\Gamma,k}/(\sum Y_{\Gamma,k}) \qquad [2.11]$$

The problem is to select from the $n\Gamma$ alternatives in such a way as to match the data as closely as possible. There are two ways formalised in the literature, one employing Lagrangian undetermined multipliers, and the other, which is effectively equivalent, to proceed with a χ^2 minimisation over admissible solutions. To use the χ^2 approach we need $\sigma_{Y,k}$, the s.d. of the data points over the replications, and

$$\chi^2 = \sum(Y_{\Gamma,k} - n\Gamma(k))^2/\sigma_{Y,k} \qquad [2.12]$$

Then

$$Q = H_{n\Gamma} - L \cdot \chi^2 \qquad [2.13]$$

where L is the undetermined multiplier. The optimum value of L, L_{opt}, is given by the gradient (∇) equation corresponding to [2.13];

$$\nabla Q = \nabla H_{n\Gamma} - L \cdot \nabla \chi^2 \qquad [2.14]$$

and

$$L_{opt} = \nabla (H_{n\Gamma})^2 \cdot (\nabla H_{n\Gamma} \nabla \chi^2)^{-1} \qquad [2.15]$$

The solution finally sought is one which satisfies $Q < \epsilon$ (of which there may be a number over an $m \times m$, lattice $\mathcal{M}$), and at the same time maximises the entropy $H_{n\Gamma}$. The transformation matrix between the Y_Γ response surface and data is the term in the model which requires iterative revision; it is the part where the effective use of prior information, and choice of starting point, should *ceteris paribus* optimise and shorten the path to convergence.

Algorithm Implementation

The choice in the examples given in the literature is usually expressed as one between models. Here there is only one model, $n\Gamma$, ($n = 2$ *or* 3 initially) but it is desired to choose between subregions of the parameter space within that model[5]. To see how many models are potentially available, it is necessary to consider what may be called the *resolution*.

Without loss of generality, consider a subregion s (within $n\Gamma$ defined over $\mathcal{M}$) as a square lattice with $s_m \times s_m$ pixels. $s \subset \mathcal{M}$. The image region is definable on a continuous response surface, so the s_m^2 pixels are a theoretical sample, t_s, in which we know, a priori, the continuity between pixels if we wish to smear local images, and conversely to unsmear them. This is the point in the argument which finds its parallels in image retrieval (Roberts, 1984).

The data matrix is $d \times d$, also square. The resolution $r_{dm} = d/m$, which is the fraction of theoretical pixels matched in the experimental design. The sampling function t_s can be varied in one or more dimensions to make the local matrix become square. Recall that the starting point of a maximum entropy problem is given as

$$\mathbf{y}_d = \mathbf{A}\mathbf{x}_m + \mathbf{b}_d \qquad [2.16]$$

---

[5] If the total region of Y_Γ is badly chosen there will not exist any subregion which is appropriate.

with $d << m$ as in Donoho *et al* (1992). The s_m^2, m^2 and d^2 matrices can be rewritten as vectors of length s_m^2, m^2 and d^2 respectively, on the real line $\Re^1$. Computationally this may be easier, but in visualising what is happening it obscures the generating psychophysics, of which it is not desired to lose sight.

It is not known what value r_{dm} has a priori, because the rate of change of the psychophysical function with stimulus parameters over the d^2 region is itself unknown; it is the purpose of an experiment to find just that. But it is possible, from the shape of the data response surface, which is equivalently the output distribution on the d^2 vector, to limit the plausible set of Y_Γ subspaces to be considered. This is the same thing as using prior beliefs about the relative probabilities of $s_m \times s_m$ $n\Gamma$ subspaces within $\mathcal{M}$; there exists a prior $p(n\Gamma)$ distribution over the Y_Γ surface, using optimal stimulus parameter settings. Regions of 2Γ for which $\partial Y/\partial a, \partial Y/\partial\lambda$ are asymptotic to zero can be excluded almost entirely. This means that the flat surface regions can be avoided.

How many submodels are there, for fixed r_{dm}, if the prior probability distribution on alternative models, $p(n\Gamma)$ is taken as flat ?

1: For each $s_m \times s_m \subset \mathcal{M}$ there are $m^2.s_m^{-2} = S_m$ positions (θ) which data can take up, the prior information is

$$H_{n\Gamma} = -\sum_{\theta=1}^{S_m} p(n\Gamma)_\theta \cdot log(p(n\Gamma)_\theta) \qquad [2.17]$$

and the maximum entropy solution over the S_m set has to reduce this $H_{n\Gamma}$, which for a flat (indifference) prior is

$$H_{n\Gamma}^0 = -\sum_{\theta=1}^{S_m} (1/S_m) \cdot log(1/S_m) \qquad [2.18]$$

and is independent of n.

The solution

$$H_{n\Gamma}' = -\sum_{\theta=1}^{S_m} p(n\Gamma|data) \cdot log(p(n\Gamma|data)) \qquad [2.19]$$

is not independent of n; a different solution may arise for each n used. The cases $a_1 - a_2 = \delta a$, λ, and a_1, a_3 fixed, $a_1 < a_2 < a_3$, λ as variables require different representation.

2: The estimation of r_{dm} within $m \times m$ $Y_{n\Gamma}$ is the problem of finding the most probable (a posteriori) $Y_{n\Gamma}$ span of $d \times d$ within $\mathcal{M}$.

So, given d and $H'_{n\Gamma}$, r_{dm} follows by minimising $H'_{n\Gamma}$. It is required to maximise the resolution within cost limits; the resolution needed to identify $m \times m$ $n\Gamma$ depends upon the local derivatives of the surface, $\nabla Y_{n\Gamma}$ within $m \times m$. For example, two surfaces $m \times m$ and $m' \times m'$ both of which incorporate and only incorporate one psychophysical gradient resembling a cumulative normal ogive are not pairwise discriminable unless resolution is asymptotic to 1. Interestingly, two chaotic (or two flat horizontal) surfaces are always indiscriminable no matter what the resolution. So the only identifiable dynamics are within those bounds of the system which limit its capacity to transmit information in the psychophysical sense. It is always possible to push r_{dm} to zero by computing finer and finer lattices in the $n\Gamma$ theoretical parameter space, with prior information on the shape of $Y_{n\Gamma}$ on $\mathcal{M}$ it is possible iteratively to recompute to make the $m \times m$ subregion spacing in $n\Gamma$ increasingly localised in terms of $\nabla Y_{n\Gamma}$.

A situation with 2 input and 2 output parameters

This example, to illustrate practical problems in using the cross-entropy, is taken from an experiment also reported by Mahar and Gregson (1994), conducted in Dr. Mahar's tactile sensory laboratory.

On the relatively hairless surface of the human forearm are placed a row of vibrotactile stimulators, each is a small plastic box with a contact surface onto the skin of about 1 cm square. The vibrators can be driven over a limited range of frequencies and amplitudes. As there is some evidence that 250 Hz is optimal for signal transmission to some of the subcutaneous receptors, frequency is fixed and amplitude of vibration taken as one independent variable. There are two vibrators in the array in this example, and the subject rests his or her arm lightly upon this array. The pairwise separation of the centres of adjacent vibrators is the second independent variable. The responses one can elicit include the total perceived intensity of sensation, encoded on some rating scale. This is done with a mouse and a p.c., using the other arm.

The total rated sensation from the array of stimulating vibrators is elicited. This is about the most basic experiment on information transmission that can be devised; it is a much simpler situation than that utilized in some prosthetic devices, which aim to replace deficient or absent hearing (or vision) with a set of cutaneous vibrators, each transmitting at varying amplitudes some part of the frequency partitioning of an auditory input.

In this very basic experiment the only patterns one can transmit are those with amplitude varying across the row vector of vibrators. The neurophysiology of even this experiment is unclear; the receptors involved are a bit equivocal, and the pathways to the brain are not unitary, and can converge and diverge. Hence, expecting a nice linear and even one-to-one mapping from stimulus parameters to the final quantified response is not very plausible.

For a single stimulus source, one vibrator on its own, the psychophysical function can be plotted, for amplitude or frequency, or for both covarying. This is what traditional psychophysics called a psychometric function. It is in fact neither a cumulative normal ogive nor a logistic function, though often assumed for tractability to be one or the other, and in nonlinear psychophysics is represented not directly but by the $\Gamma : a \mapsto Y$ input-output mappings. Given a set of fixed levels of amplitude (or frequency) the information transmission of that created sensory channel can be measured. Like in olfaction, it is not much. So we have two inputs, of equal value ($U_1 = U_2$) and the subject is asked to judge and report the total intensity I ($= I_1 \odot I_2$) on a bounded (continuous) scale, by adjusting a pointer on a computer screen. The precise form of the operation $\odot$ which is used by the observer to coalesce inputs is unknown. This is modelled as 2Γ **Case 2**,(see chapter 1), with $a_1 = a_2 = a$, and λ as the parameters defining the output space in $Y_{tot}(\mathrm{Re})$; the two empirical variables are input amplitude, $v_1 = v_2 = v$ of two 1 cm $\times$ 1 cm vibrators, and vibrator spatial separation s, varying between about 3 and 7 cm.

The response is perceived total sensory intensity, I, so $a \Leftrightarrow v$, $\lambda \Leftrightarrow s$.

Constructing the Entropy Surface

The first step is to construct a map of the Boltzmann cross-entropy over the parameter spaces of the theory and data combined. The data have two parameters v, s, and it is assumed that $v = f(a)$ and $s = f(\lambda)$ are the correspondences that we seek to establish between theory $Y_{tot}|a, \lambda$ and data $I|v, s$.

The data only exist for a subset of the possible theoretical predictions, and both the total theory surface, generated by spline interpolation from a lattice of pixels, and the results from a factorial design on the parameters, are taken as square matrices here.

That is, the theory is definable as a dense rectilinear lattice of parameter values, the density of coverage being arbitrary, and the observed data are a corresponding but very sparse lattice, which is to be mapped within the area of the theoretical lattice. In the current example the theory lattice is taken as 25×25, the data lattices are 5×5, there are possible replications of the data, introducing error variance due to imperfect reliability, but not of the theory. It is convenient to redefine [2.3] in the form of a bivariate summation which is H_B, as in [2.20]. Because of the a *priori* defined one-to-one correspondence between parameters, a notation can be used which reflects the situation succintly.

$$H_B = -\sum_{v,(a)}^{n} \sum_{s,(\lambda)}^{n} p_{v,s} log_e(p_{v,s}/m_{a,\lambda}) \qquad [2.20]$$

The theoretical data set Y_{tot} generated on a 25×25 matrix becomes, by affine rescaling

$$m_{a,\lambda} = \alpha_\Gamma + \beta_\Gamma \cdot Y_{tot} \qquad [2.21]$$

and the experimental observations $I_{v,s}$ on a 5×5 matrix become

$$p_{v,s} = \alpha_{data} + \beta_{data} \cdot I_{v,s} \qquad [2.22]$$

Note that I is bounded $0 < I < 100$ in the experimental setup by a scoring convention, and Y_{tot} is bounded by the range over which simulation is performed. In fact this latter range is adjusted by the two parameters a, λ acting jointly for each subject to trap the region which minimises H_B. It is important to note that this is not a psychophysical scaling experiment; the units of both theory and data are arbitrary, the objective is to estimate the topology of the response surface which best represents the nonlinear system involved, as a compromise between the theory as starting point and the noisy data as observations.

The cells in Tables 2.1 and 2.2 are not v, s coordinates in the data, they are a, λ locations determining where the $p_{v,s}$ grid is superimposed upon a subset of the theory lattice. The $p_{v,s}$ grid is always spaced regularly (with a second-order correction based on the physical calibration of the vibrator amplitudes) on the square lattice of a, λ values; the starting point of the minimum cell in a, λ coordinates can take 25 possible locations, on the square with corners $1, 1; 1, 5; 5, 1; 5, 5$; the matrices in Tables 2.1 and 2.2 correspond therefore to these possible positions of superimposition.

The rescaled variables p, m are each made to lie in the range $0, 1$. The original (5×5)data have to be located by superimposition on the $N \times N$ (that is, 25×25) theory matrix. That is, $n << N$, and the comparison minimising H_B is found by moving the sparse data matrix over the dense theory matrix, preserving parallelism of the parameter axes. That is, linear translations or dilations are used, but not rotations. This results usually in the identification of a local minimum for H_B, but this need not be unique.

There is, however, another refinement in considering psychophysical data which is not usually met in simple regression models. Working towards a solution in the parameter space both minimises H_B, and in the regression equation

$$p_{v,s} = \alpha + \beta \cdot m_{a,\lambda} \quad [2.23]$$

induces a solution where $\alpha \to 0$, and $\beta \to 1$, and the percent variance taken up by this regression of 25 p means on 25 m means is maximised. These criteria cannot necessarily all be satisfied at the same time; the best regression is then sought within the subset of a, λ ranges which minimises H_B. But the data means are not all of equal worth, those with larger associated inter-trial variance (i.e., low reliability) should be proportionally discounted. There are various ways in which this might be done; for example, starting from the Fisherian definition of statistical information as inverse variance, and avoiding computational problems for cells with zero data variance, (that is, with complete response stereotypy) Defining

$$c_{v,s} = (\sigma^2_{v,s} + 1.0)^{-1} \quad [2.24]$$

and using this c as what may be called an F-weight on terms in H_B, gives

$$_wH_B = -\sum_{v,s}^{n,n}\sum_{a,\lambda}^{n,n} p_{v,s} log_e(p_{v,s}/m_{a,\lambda})c_{v,s} \quad [2.25]$$

In each data block for a subject treated separately (pooling across subjects is meaningless and inadmissible in this approach), the a and λ ranges have been adjusted so that the $min(H_B)$ and $min(_wH_B)$ values (which usually correlate highly) lie in the centre of the matrix of Boltzmann values.

We examine data from one subject on 125 trials in graphing and in numerical detail, to show the procedure, and show collateral data from three other subjects only in graphical form.

In Table 2.1, the a, λ values in the centre (3,3) location are generated by the combination of the vectors

$$a = 2.05 + 3 \times .05, 2.05 + 8 \times .05, ..., 2.05 + 23 \times .05$$

$$\lambda = 0.45 + 3 \times .005, 0.45 + 8 \times .005, ..., 0.45 + 23 \times .005$$

Table 2.1: data for subject s11

η = run within nonlinear trajectory is 10,
max and min mean data response: 0.9200 0.0000,
max and min 2Γ **Case 2** theory: 1.3128 0.1606.

Matrix of Boltzmann Values, Unweighted and Weighted
for 5×5 positions of data lattice superposition on theory

1	-2.91786	-1.36620	-0.10385	0.71717	1.57573
	-2.81563	-1.30872	-0.08246	0.71729	1.55214
2	-2.73683	-1.23941	-0.01144	0.62723	1.68099
	-2.64006	-1.18573	0.00720	0.63150	1.65386
3	-2.56728	-1.11774	0.07818	0.78416	1.77141
	-2.47561	-1.06770	0.09415	0.78305	1.74130
4	-2.40709	-1.00067	0.16519	0.90594	1.85272
	-2.32023	-0.95412	0.17856	0.90079	1.81997
5	-2.25479	-0.88776	0.24973	1.01165	1.92765
	-2.17249	-0.84459	0.26059	1.00306	1.89249

In each cell the unweighted value is given above the weighted value.

Combining Theory and Data to get an Estimated Response Surface

Considering only the Y_{tot} theory and I data values for the optimal case with minimal entropies over the 5×5 $p \times p$ matrix, these 5×5 pixels are replaced within the m surface with a weighted compromise between theory and response, and the remaining pixels are left at their theoretical values. Again the inverse variance [2.24] is used[6],

$$x_{a(v),\lambda(s)} = c \cdot p_{v,s} + (1 - c) \cdot m_{a,\lambda} \qquad [2.26]$$

Note that the data suggest a very nonlinear dependence of I on v, s, and that they are markedly heteroscedastic. The $\sigma^2_{v,s}$ terms are here based on 5 replications of each stimulus condition, and extremely high or low intensities can exhibit zero response variance. In addition, a rough approximation to Ekman's Law holds in that the larger mean responses have larger response variances. The x values of [2.26] are put into the m surface as local replacements, and the

[6] The weighted average here resembles the Bayesian treatment of contaminations (Moreno and Cano, 1991). When the combined $2\Gamma, x$ matrix is constructed it can be used as a prior distribution on subsequent experiments.

Table 2.2: subject is s11
details of best fit location for data
on $a(v)$, $\lambda(s)$ coordinates

top left coordinates within 2Γ 25x25 array are 3,3
first line is mean data rescaled
second line is theoretical 2Γ values
third line is standardized data error

1	0.108696	0.108696	0.010000	0.010000	0.021739
1	0.032689	0.037092	0.041573	0.045215	0.048909
1	0.34803	0.32787	> 3	> 3	-0.61085
2	0.260870	0.282609	0.108696	0.239130	0.260870
2	0.077409	0.104554	0.134488	0.160293	0.187555
2	0.96296	0.72669	-0.36981	0.95746	0.32611
3	0.695652	0.543478	0.565217	0.500000	0.478261
3	0.592038	0.616456	0.638848	0.655374	0.670738
3	0.39296	-0.39904	-0.50709	-1.38915	-1.46443
4	0.869565	0.717391	0.869565	0.673913	0.782609
4	0.866429	0.870260	0.874006	0.876947	0.879842
4	0.01370	-0.93394	-0.02428	-0.83527	-1.18087
5	1.000000	0.760870	0.804348	0.932609	1.000000
5	0.965132	0.966832	0.968544	0.969922	0.971306
5	0.27292	-1.21591	-0.92448	-0.42902	0.64513

Boltzmann unweighted matrix sum of 0.0782
Boltzmann weighted matrix sum of 0.0941
raw regression intercept, slope, var pct 0.027 1.106 92.397
weighted regression intercept, slope, var pct 0.027 1.105 92.371

shape of the resultant contour maps without and with replacements are then compared. All the graphs show there is disparity with the theoretical response surface for low separation s and high amplitude v inputs, and better matching elsewhere.

Examples of the response surface maps for four subjects are shown in Figures 2.1 through 2.4. The first two cases are from subjects who made a

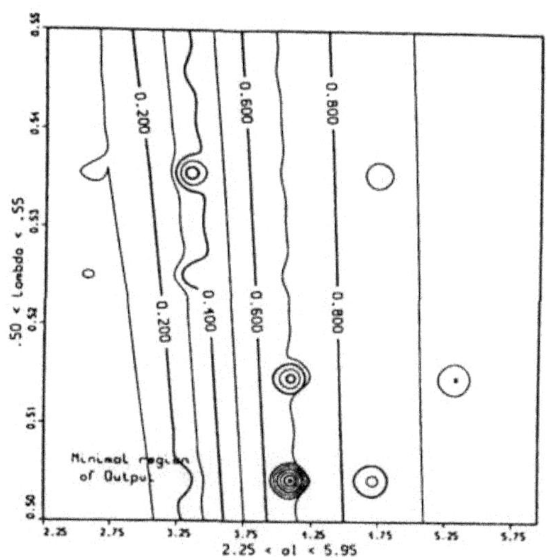

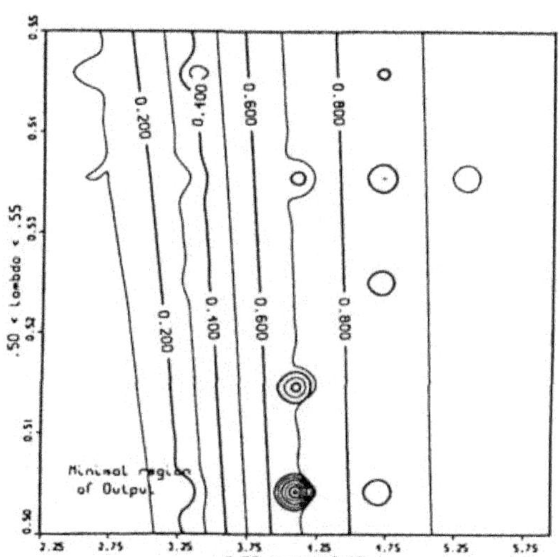

Figures 2.1 and 2.2: with single response per trial

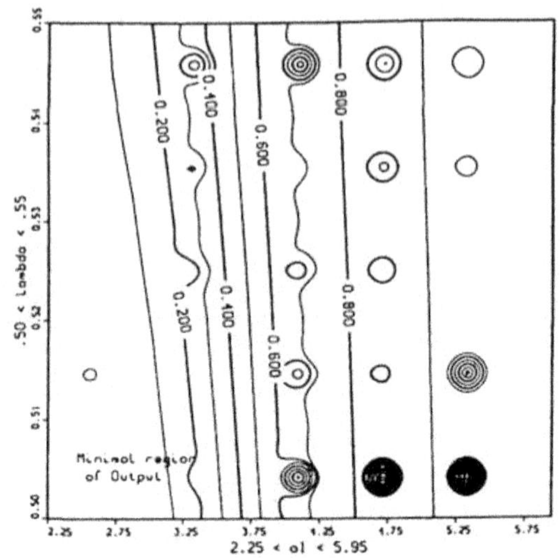

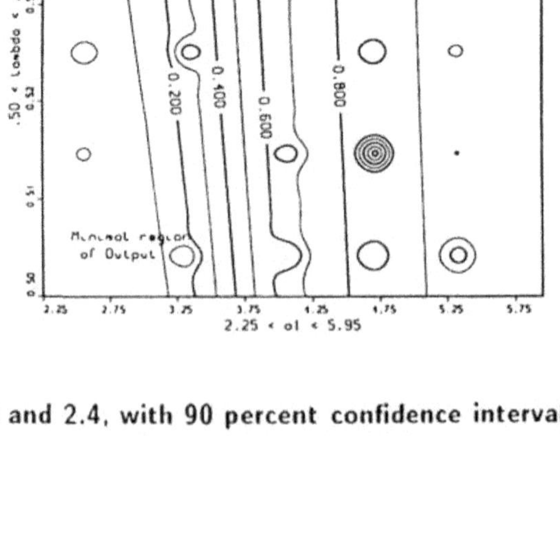

Figures 2.3 and 2.4, with 90 percent confidence interval midpoints

single response on each trial, the second two are from data where the subjects gave the 90% confidence limits on their judgments. The numerical values of the original $Y_{tot}|2\Gamma$ and the x, m surface are obviously different because of the units used, it is the shapes only which are being compared. When a data point p fails to fall onto the theoretical response surface m, it produces a local concentric contour pattern. When it fits it vanishes. There are at least six reasons why failure to lie on the surface could arise:

(a) the measures chosen for the physical stimulus values are wrongly defined,

(b) the model fails to predict secondary nonlinearities, often at extremal values,

(c) insufficient data are available to stabilize mean responses,

(d) there are end effects from the use of a bounded response scale,

(e) the subject is at or above a bifurcation in his or her dynamics and consequently responses are bimodally distributed,

(f) the representation of $I_1 \odot I_2$, the sensory pooling of two equal inputs, is not properly modelled by the additivity of Y into Y_{tot}.

For (b,f) the situation is most interesting because it is one where psychophysiological data can be drawn on to suggest why, in all the graphs examined, there are departures from the response surface at high input a and low spatial separation (this result holds for other data not shown here).

Microphysiological Correlates

The innervation of the subcutaneous layer involves four types of receptors; RA, PC, SA I and SA II. The work of Johansson, Johansson and Vallbo, and Vallbo and Johansson (1976, reported together) suggests that these four types have very distinctive functional characteristics (p. 169); two variables have to be considered, sensitivity to sustained indentation and sensitivity to remote stimuli. These together constitute a basis for the classification of the mechanisms. For spatial discrimination and accurate localization, the RA and SA I units have small and well delineated receptor fields. But the other two types, PC and SA II (with which we are most probably concerned here on the forearm hairless surface) appear "to analyze other aspects of the mechanical events. They are are less suitable for spatial discrimination but could monitor mechanical vibrations travelling through the tissue and stretching the skin". This in itself is not linear when we have two or more inputs as arranged in our experiment; "the receptor field of any unit is larger the higher the indentation amplitude" (p. 176). The recuitment of adjacent units is marked; a doubling of amplitude can result in a tenfold increase in recruitment. It also appears that it is dangerous to generalize too readily from results on one skin area to another; "even the smallest variation of a suprathreshold stimulation, e.g. variations in location and in the magnitude, will appreciably change this pattern (of neural activity) "(p. 192).

Multiple Mapping at Response Bounds

The mean data values may be relaxed in the sense that they are replaceable by their 90% confidence bounds. That is,

$$if \quad p_{v,s} - 1.6 \cdot \sigma_p < m_{a,\lambda} < p_{v,s} + 1.6 \cdot \sigma_p, \qquad then \ p^*_{v,s} = m_{a,\lambda} \qquad [2.27]$$

and the p^*, m surface is mapped instead of the p, m surface. This eliminates only those disparities which are associated with response imprecision, where that imprecision is is based on intertrial variance. Using [2.26,2.27] means that poor data should make negligible revisions to theoretical predictions, except where the poor data are stereotyped responses with zero intertrial variance. If the extreme judgments are only associated each with a single (v, s) location in the experiment, and the theoretical Y surface has only one maximum and one minimum, then the stereotypy can be modelled. Otherwise it cannot.

However, there is another objection to using [2.27], it relies too readily on an uncritical acceptance of the Central Limit Theorem, and ignores the actual data. Such a stance is not good Bayesian procedure. Assuming that the intertrial variance of responses is the only appropriate measure of imprecision, when that imprecision is itself directly accessible on a single trial and may be a variable, is not compatable with some current statistical thinking (Walley, 1991) which rests precisely on personalist imprecision. Instead of collecting one response on a trial, we may collect the lower and upper bounds of a confidence region, spanning some subjective probability region. Call these bounds $p_l(v, s)$ and $p_u(v, s)$; then a condition for coherence is that

$$F_c : \quad \forall\, p_{v,s}, \ sup(min\ p_l(v, s)) < inf(max\ p_u(v, s)) \qquad [2.28]$$

and only where the condition F_c is satisfied is it meaningful to attempt a representation in 2Γ.

This [2.28] is a very strong constraint on data acceptability, it means that only regions where the upper limit of the lower bound, across replications, never crosses the corresponding lower limit of the upper bound, are meaningful. In other words, when using dual responses we exclude by definition the sort of distributional overlap that generates false positives and false negatives in signal detection models. If violations of F_c are admitted then the theoretical $n\Gamma$ surfaces have to be multivalued in some subregions; this can be done in $n\Gamma$ theory by pushing the a, λ values up to generate multiple Y_{obs} (for small η variations) beyond the onset of limit cycling, but this is at the price of using parameter values that imply very high stimulus levels. If the violation of F_c arises in, and only in, regions where v is high and s is small then such modification is plausible, but not otherwise.

The use of F_c and p_l, p_u means that two surfaces can be generated over a consistent response region, and that the Γ surface which satisfies minimum cross-entropy is sandwiched between the $p_l(v,s)$ and $p_u(v,s)$ surfaces without touching either at any point $a, \lambda \Leftrightarrow v, s$ unless the p_l and p_u surfaces locally coincide at that point. This procedure obviously entails a different experiment with different instructions to subjects.

Combining Dual Responding with Theoretical Output Variance

The 2Γ theory can also be used (Gregson, 1992a, Chapter 8) to predict the local variance in Y_{tot} as a function of the gradient ∇Y of the response surface, which then gives an additional check on the model by comparing the topology of the $c_{v,s}$ over the data matrix with that of ∇Y_{tot} in a local neighbourhood of the a, λ space, which it should resemble for corresponding locations $a(v), \lambda(s)$.

In effect, there can be two theoretical surfaces, $n\Gamma$ and $\nabla n\Gamma$; the first has been used for the m, p matchings, the second can be used in the following sense, by defining what may be called a *trapping relationship*.

If $\nabla n\Gamma$ is used to estimate a local ∇Y_{obs}, then for some constant ξ,

$$\forall v, s, a, \lambda \; \exists \xi;$$

$$Y_{a,\lambda} - \xi \cdot \nabla Y_{a,\lambda} \leq inf(min \; p_l(v,s)) \qquad [2.29]$$

$$Y_{a,\lambda} + \xi \cdot \nabla Y_{a,\lambda} \geq sup(max \; p_u(v,s)) \qquad [2.30]$$

In graphical terms, we seek the minimum value, in a ξ confidence region, associated with $Y|2\Gamma, a, \lambda$, which will enclose at all points the lowest and uppermost values of the dual responses. Because of the way the responses are generated, $min\xi$ is also a function of the observer and not only of the theory; the desideratum is to minimise ξ over the region which satisfies F_c, which is to say the trapping should be in a minimum $n\Gamma$ volume.

Choosing between models

Obviously there are competing theories about the shapes of response surfaces, apart from the Boltzmann [2.3] optimisation choices created here for one theory, 2Γ. To decide on the posterior probability of a theory after an experiment, and at the same time utilise the cross-entropy measures, heuristically it is suggested that an extension of Bayesian inference could be tried. Using Bayes' theorem written with likelihoods in its usual form, for example, with the chosen competing theories 2Γ nonlinear psychophysics, M_1^2 city block metric space scaling, and M_2^2 Euclidean metric space scaling, gives;

$$p(2\Gamma|Data) = \frac{p(Data|2\Gamma)p(2\Gamma)}{p(Data|2\Gamma)p(2\Gamma) + p(Data|M_1^2)p(M_1^2) + p(Data|M_2^2)p(M_2^2)} \quad [2.31]$$

and so on for the posterior probabilities on M_1^2 and M_2^2.

However, there is no direct conditional probability function, but a weight-of-evidence function is needed in its place to substitute in [2.31].

The cross-entropy of the fitted similarity response surface to the data is $H_B|2\Gamma$, or $H_B|M_1^2$, or $H_B|M_2^2$. Call these $H_{B|X}$ for short; the maximum disparity value of the cross-entropy in the data lattice is ${}_{max}H_{B|X}$. This represents a worse case, random m_i values would serve as an analogue of null hypotheses in Neyman-Pearson inference, and would constitute a maximum entropy prior. Two other alternatives are possible; a completely uniform prior, which represents zero information like an indifference prior, or the mirror image of the data, (hills for valleys and *vice versa*) which should produce an absolute maximum departure from any model with prior psychological plausibility.

Then

$$\Lambda_{B|X} = \frac{|{}_{max}H_B| - |H_{B|X}|}{|{}_{max}H_B|} \quad [2.32]$$

where $0 \leq \Lambda_{B|X} \leq 1$ is a pseudolikelihood. Substituting in [15] we get

$$p(2\Gamma|Data) = \frac{\Lambda_{B|2\Gamma}p(2\Gamma)}{\Lambda_{B|2\Gamma})p(2\Gamma) + \Lambda_{B|M_1^2})p(M_1^2) + \Lambda_{B|M_2^2})p(M_2^2)} \quad [2.33]$$

and so on for the posterior probabilities.

The choice between models is simply to pick the one with maximum posterior probability, given the data. A posterior likelihood ratio for two competing models, for which we have no prior information, is given by

$$\frac{p(S_\Gamma|RespSurf)}{p(M|Response\ Surface)} = \frac{|{}_{max}H_B| - |H_{B|\Gamma}|}{|{}_{max}H_B| - |H_{B|M}|} \quad [2.34]$$

and the r.h.s of [2.34] can be thought of as the complement of cross-entropy. However, the choice of prior probabilities on the three models is a matter of importance, in which consideration of previous data is involved.

Comments

The approach here requires that a complete theory of the topology of the response surface is available, but does not impose any restrictions on the linearity of the process generating the responses. Geldard (1975, p 21), writing on tactile sensation, comments that "The physiological organism... has dimensions of its own to contribute, makes its own transformations, and creates its own nonlinear functional relationships in the devious paths from peripheral receptor processes to final response mechanisms". It is precisely some of those nonlinearities that we aim to model by employing 2Γ dynamics here.

The 2Γ model was selected because it is already known to have utility in reproducing a diversity of psychophysical situations; tactile sensitivity is an addition to phenomena previously modelled (Gregson, 1989, 1992a, Gregson and Britton, 1991, Gregson and Harvey, 1992, Gregson and Wiggins, 1990). The Boltzmann approach is new to this study; the notion that we might locally relax the fit because of the intrinsic imprecisions of psychophysical response processes is in the spirit of Walley (1991) where probabilities are replaced by upper and lower bounds or what he terms *previsions*. It is notable that the idea of substituting two bounds with imprecision for one estimate with precision is a statistical analogue of the double estimates in the psychophysical method of limits; however the tradition in psychophysics averages out and thereby destroys some of the informative properties of the confidence intervals on a response. It is not necessary to implement the method of limits in such a fashion within nonlinear psychophysics (Gregson, 1992b). Restricting mapping to the positive F_c region as in [2.28] is a way of restricting a model to data regions which exhibit stationarity in the response generating process. As the 2Γ model used here is stationary in its parameters it can only legitimately be applied under such a constraint. However, it can be extended to higher dimensionality systems, when we deal with response hypersurfaces and sections through them.

The only identifiable dynamics are those within the bounds of the system which limit its capacity to transmit information in the psychophysical sense. This suggests that an extension of minimum cross-entropy theory to incorporate fuzzy set properties of data, which has been anticipated by Jumarie (1984), is also worth consideration. The responses of human observers do not have tidy properties which map into the set theoretical bases of probability theory, the fuzziness is thought of as the result of coupling effects with semantics, and 'subjective entropy' then replaces entropy as the basis of information. It is always possible to push the resolution of data on theory (here $5/25 = .2$) to zero by computing finer and finer lattices in the 2Γ theoretical parameter space. Obviously this trivializes the exercise if done; but it is pertinent to point out that if a psychophysical process is very nonlinear then a sparse factorial design can approximately fit and tell us nothing about the underlying dynamics. With prior information on the shape of $Y_{2\Gamma}$ on $\mathcal{M}$ it is possible iteratively to recompute to make the $n \times n$ subregion spacing in 2Γ increasingly localised in terms of $\nabla Y_{2\Gamma}$. In this context, the number of parameters in $n\Gamma$ is minimal, the ambiguities reside in the intrinsic lack of defined metrics on the bounded response scales, and hence on the mappings $Stimulus \mapsto a, \quad Y \mapsto Response.$ These have been taken as affine, which is itself a simplest form. In contrast with statistical modelling, the nonlinear psychophysics is deterministic, the only stochastic components here are in the response selection process, as shown in the lack of reliability, or perhaps paradoxically at some scale anchor points, in the complete response stereotypy.

The use of [2.3,2.22] in this situation effectively ignores the points $m_\omega \in \{m \times m - n \times n\}$, that is, the parts of the 2Γ surface for which no corresponding data have been collected, but could there be collected in another analogous experiment by interpolation at any $\{v, s\}$ which are possible to implement. This is, in fact, a strictly bounded set; in our particular case the human forearm has no infinite extensions. The set $\{m_\omega\}$ is not used here to calculate the value of $minH_B$, but the data are located on $m \times m$ by computing H_B on a series of $n \times n$ lattices each using subsets for which H_B is found not to be minimal. But if ∇Y is computed, the points in $\{m_\omega\}$ adjacent to those in $\{m\}$ are also used, but the resulting ∇Y lattice is on $n \times n$, and hence both the upper and lower ξ bounds in [2.26,2.27] are again on $n \times n$.

The question can be put, will the location of $\{m\}$ which minimises ξ, given 2Γ and H_B, be the same as that whch minimises H_B or ${}_wH_B$? The entropies of the ∇Y surfaces are calculable, as is the entropy of a $var(p)$ surface, and cross-entropies between those surfaces can consequently be derived. Only if, at all a, λ, ∇Y is proportional to Y will $\xi \nabla Y$ leave H_B unchanged and independent of ξ; this condition is obviously not globally true in a nonlinear system. If, therefore, we reject the exclusive use of single-valued responses in a psychophysical task, the resultant estimation of the response surface will be different at least in some critical regions of the sensory channel's parameter space where the psychophysical gradient is steep. That is, breakdown of Weber's law is reflected in the different estimates of H_B generated from Y and from ∇Y.

These idiosyncratic features of psychological data need a treatment that does not derive readily from the probability distributions used in some statistical modelling; one possible way out may be to replace the continuous scoring of responses by category scale partitioning, to admit thereby a multinomial encoding. This would open the way to employing the logistic-normal approximation to the Dirichlet distribution (Block and Watson, 1967, Aitchison and Shen, 1980) as a treatment of the final mapping $Y \mapsto Response$ (the second affine assumption following [1.1]). But in doing this, we then create for every pixel on the (a, λ) surface an associated logistic-normal distribution, for which some ancillary assumptions of coherence over the whole $Y_{tot}|2\Gamma(a, \lambda)$ model would be needed. This becomes a substitution of a task in massively parallel computing for the two simplistic mapping assumptions used here.

3 Properties of Nonlinear Dynamics underlying the generation of Cascades and Fields

"the outward forms of our mathematics are not absolutely relevant from the point of view of evaluating what the mathematical or logical language truly used by the central nervous system is...the system...cannot fail to differ considerably from what we consciously and explicitly consider as mathematics."

J. von Neumann (1954) *The Computer and the Brain.* Yale University Press, p. 82.

Restabilisation Trajectories

Nonlinear psychophysical models represent the action of a stimulus input as something extended in real time, and as a perturbation on the dynamic equilibrium of the continuous system which the input destabilizes in a transient manner. This form of model has some resemblance to physiological accounts of mass action in neural networks mediating olfactory sensation (Skara and Freeman, 1988). However, so far no specific attention has been given to the problem of how we might represent the way in which the system gets back to its stability after the trajectory consequent upon a stimulus input has run its course. Any model that is in the widest sense cybernetic has to make provision for its dynamics to be in some degree self-correcting. The problem of economically representing this restabilization, with minimum parameter elaboration, and by using essentially the same sort of nonlinear dynamics, is addressed here. The problem links into other issues in psychophysics, such as adaptation level effects, and the possible existence of refractory periods between successive stimuli being processed serially in a single channel. It is expedient to emphasise here that the model does not involve any representation of memory or information storage; all the dynamics represented are in a stage that precedes input even to short-term memory. The only continuously represented parameters are those which define the baseline (Y_s) of the system's dynamics; these are state variables and not internalised responses.

Starting again from the single channel model, Γ, [1.1] involving the mappings

$$U \mapsto a, \quad \Gamma_{a,e,\eta}(Y(Re, Im)) \mapsto Y(Re)$$

for given starting values of Y, and fixed values of a, e, the notion underlying this representation of a stimulus input, and its subsequent evolution within the organism in real time, is that Y_s, a complex number, is the stable baseline of activity of the internal variable which sufficiently characterizes the system, and the transient destabilization is induced by the operation of the input U onto a, pushing a away from the current value associated with Y_s, which we will call a_s. This destabilization is explicit but not treated *in extensio* in its implications in Gregson (1988,1992a). No discussion has previously appeared on how precisely such a system gets itself back to stability. Apparently this metatheoretical problem arises in any psychophysical model, accumulator (Vickers, 1979), random walk to absorbing barrier (Heath 1981,1982), or attractor, which allows a process to meander in real time.

Inspection of the trajectories of the Γ evolution indicates that provided the system stays within the region of its parameters which avoids bifurcation dynamics and consequent limit cycling or chaos, then for some a the Y(Re) trajectories run to asymptote above the initial Y_0, have positive slope and are negatively accelerated, whereas others run to asymptote below their Y_0 (which need not be equal to Y_s), have negative slope and are positively accelerated. All trajectories of this sort run onto a point attractor, where the gradient of the trajectory, $dY/dj, = 0$. The Y(Im) trajectories run onto the real line and eventually vanish (Campbell and Gregson, 1990).

It is therefore desired to find the a_0, for any given Y_0, e which ensures that $Y_0 = Y_\eta$. This may be approached as a computational problem; as the trajectory is in discrete and not in continuous time, by computing the $\Delta^1 Y$ trajectory, and finding its zero values. This is analogous to finding the roots of an equation by the bisection method, to arbitrary accuracy (Press, Flannery, Teukowsky and Vetterling, 1986).

Computations

The initial Y_0 is fixed, and for each a is found the asymptotic value of $Y_\eta = Y^*$ for which $\Delta^1 Y_\eta = 0$. This is $f(a)$. Using Y^* as a new Y_0 to generate a new Y_η^*, then an a_0 such that $Y_\eta^* = Y_0$ is found. The duration of the Γ trajectory with a_0, Y^* is the refractory period of the system so long as $Y^* > Y_0$.

If we start with a given Y_s as the system's stable baseline level of activity, we can ask, is there an a_s (an internal restabilization gain) such that the process remains at its baseline indefinitely if not given sensory inputs ?

From [1.1], it is easier to put the question in reverse, and then consider the trajectory when

$$\forall\, j > \omega : \quad Y_{j+1} = Y_j \tag{3.1}$$

in order to ask what $Y_s | a_s, e_s$ will satisfy [3.1] ?

However this leads into the possibility that no algebraic solution exists, because the cubic obtained by expanding [1.1] where [3.1] holds can produce (for some coefficients) the irreducible case (see Ferrar, 1950, chapter XXI).

It is relatively simple to solve [3.1] numerically in order to find a real root or roots for Y, and generate the family of curves, each relating Y_s to a_s for a fixed e_s. This is the same problem as asking, given a first Y_η, what a_0 will return the process to Y_0 when $Y_\eta \neq Y_0$? Graphs of some trajectories, for $\eta = 20$ and $\omega = 20$ are given, for a variety of Y_0 and a values, in all cases $e_0 = 10^{-5}$.

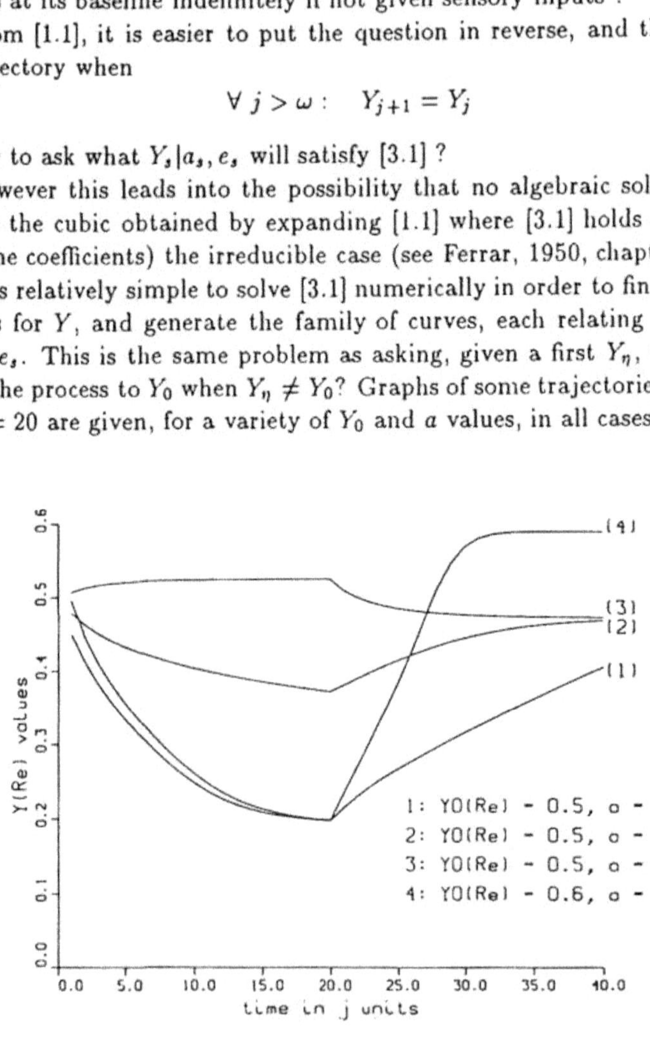

Figure 3.1: Restabilisation Trajectories; middle baseline

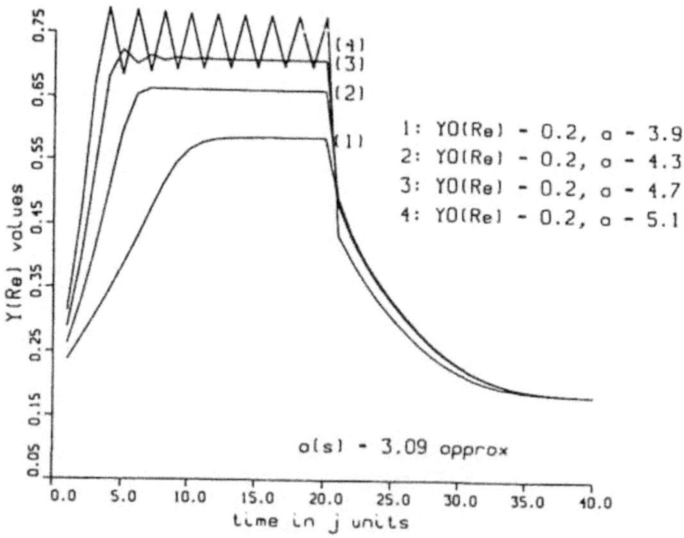

Figure 3.2: Restabilisation Trajectories; very low baseline

It is noted that the shapes of these curves could in some cases be readily and perhaps misleadingly simulated with exponential decay and growth curves; experimental data on psychological adaptation (Ekman, Berglund, Berglund and Lindvall, 1967), or the growth and decay of iconic images (Loftus, Duncan and Gehrig, 1992) provide examples.

The curves in Figures 3.1 through 3.3 are created by using fixed η and ω; this is an oversimplification because the trajectory may terminate not at a fixed j value ($= \eta$), but with a probability distribution over the j steps. This distribution is derived in Gregson (1992a) from the non-constant Jacobian of the $n\Gamma$ trajectories. Examples of trajectories generated from very low, and very high, baselines are shown in Figures 3.2 and 3.3; one of the trajectories is moving into chaos at the time it is restabilized. but this does not impede the restabilization. It simply means that the starting point of the restabilization trajectory can be itself unstable and not a single-valued asymptote. However, by the peculiarities of the algebra we run into trouble if Y_s is set very high; then the Γ trajectories, even for high a, run to stability at $Y << Y_s$ very rapidly, which induces somewhat paradoxical dynamics of restabilization. $Y_s \leq .7$ appears to be an ad hoc

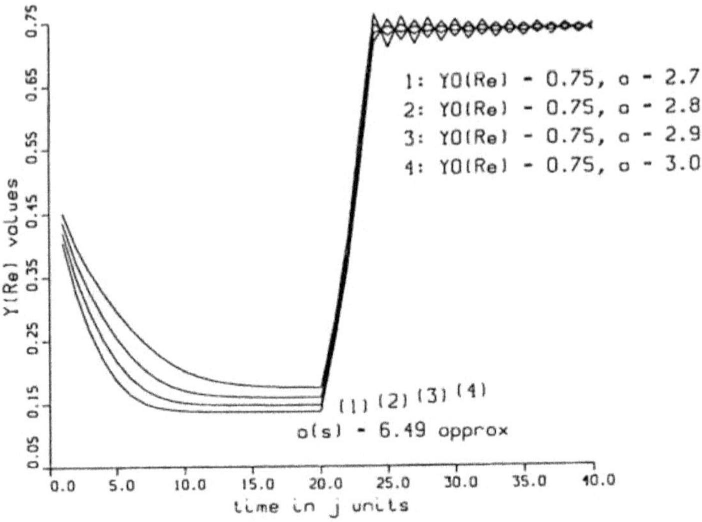

Figure 3.3: Restabilisation Trajectories; high baseline

condition to avoid this. Empirically this might mean that in a Γ system inputs below (or even slightly above) a high baseline have degenerate consequences, which is psychologically a nice result if it can be sustained. Inputting signals to a high baseline results in their rapid deletion, but apparently at the price of creating some local transient second-order oscillations in the baseline Y_s itself.

If the initial trajectory terminates before the $\Delta^1 Y(\text{Re})$ asymptote then the restabilization trajectory has less distance in the Y direction to move before regaining baseline values. After asymptote it does not matter; the restabilization time is then independent of the initial η.

Adaptation Level Dynamics

The asymptotic Y value to which the Γ trajectory returns is independent of Y_0 whether or not the dynamics are convergent on a point attractor, though the time to return (ω steps of the recursion) is not. An a_η value, such that [3.1] holds for the first Y_η may exist; that a_η will be the value of the destabilization that moved Y_0 to Y_η, and is also the operation of a baseline shift, or complete adaptation of the baseline.

Partial shift to baseline (with incomplete recovery to a fraction κ of the displacement $Y_0 \Rightarrow Y_\eta$) can occur. Suppose there is such partial recovery, then the situation reduces to a cycle of steps J, $J = 1, 2, ...$

$$Y_{0,J}|a_J \rightarrow Y_{\eta,J}|a_J \rightarrow \Gamma(1-\kappa)a_s \rightarrow Y_{\omega,J} = Y_{0,J+1} \rightarrow Y_{J+1}|a_{J+1} \rightarrow$$
$$Y_{\eta,J+1}|a_{J+1} \rightarrow \Gamma(1-\kappa)a_s \rightarrow \; etc \qquad [3.2]$$

where ω in j units is the interval between one internal response and the next stimulus, that is, the intertrial interval less the process internal latency. The J steps are thus each $\eta + \omega$ in duration. This cycle is a special case of cascading; the example given by Gregson (1988), is one where the degree of restabilization corresponds to $\kappa = 1$ in [3.2].

The concept of an adaptation level for sensory judgments was made central to psychophysics by Helson (1964), and a variant of that concept can be incorporated in the current framework. In effect it is aimed to replace the notion of adaptation level, as a separate mechanism with its own parameters, as it were linking sensation to perception, with the intrinsic dynamics of restabilization.

Helson used formulae which involved the products of three terms; the current stimulus, the context or standard, and the adaptation level already established in the process[1]. Responses were often on category rating scales.

---

[1] Helson (1964, p. 129) writes an equation as

$$logAL = logK + \left[\frac{\left[k_1 \sum_J^n logX_J/n\right] + k_2 logC}{k_1 + k_2}\right] \qquad [3.2A]$$

which means that the total record of a series of X_J presentations is implicated. This assumption leads to trouble (Gregson, 1983, comments thereon); the series $\{X_J\}$ should be weighted and truncated. Putting this objection temporarily aside, it can be shown that the [3.2A] expression involves not only the stimulus X and context C, but also the previous AL, by rewriting [3.2A] as a recursion

$$logAL_{J+1} = f(logAL_J, X_{J+1}). \qquad [3.2B]$$

The constants k_1, k_2, K take values partly determined by the scale units in which stimuli and responses are measured. Helson notes (p. 129) that his AL is "changing at every moment of time, and values assigned to AL must be regarded as means". As [3.2A] involves a context dimension C as well as the series $\{X_J\}$ it needs rewriting as 2Γ in the present context if its logic is to be preserved.

The weights of the terms in such equations represented the relative contribution of the terms to determining a current response. Such equations are *ad hoc* in the sense that they do not derive from a model of the underlying psychophysics in real time, as does the present development from Γ theory. Here only one new parameter κ is added, which is used to perform the same task, within the algebra, as the three weights in a Helson equation. As those weights are relative values they possibly introduced only two new free parameters. It is therefore of interest to see if predictions about the shape of adaptation curves over a series of trials can be derived from the present model. Responses can be predicted either on continuous scales or on category scales.

If the cycle [3.2] generates a convergent series in $Y_J(\mathrm{Re})$ for fixed $a_J = const$, then where it converges depends upon a_J and κ; they are in a Γ process the *joint* parameters determining both the adaptation level and the rate at which it is approached. The asymptotic Y_∞ never fully returns to Y_0 if $\kappa > 0$.

In [3.2], if $\kappa = 0$ then

$$Y_{\eta,J}|a_J \rightarrow \Gamma(a_s, e) \rightarrow Y_s|a_{J+1} \rightarrow Y_{\eta,J+1}|a_{J+1}. \qquad [3.3]$$

If $\kappa > 0$, and we employ an i.i.d. stimulus series of a_J (the randomness generated in the environment U) we have a situation that resembles $\Gamma V1$ (Gregson, 1988) but is not exactly the same. In $\Gamma V1$ the difference of inputs in successive trials, $U_{J+1} - U_J$ has a constraint imposed, so that

$$|\Delta^1 a_J, a_{J+1}| < c, \qquad [3.4]$$

and $\forall J : Y_0 = Y_s$. Here, the $\Gamma(\kappa, a_s)$ process runs through $\eta + \omega$ to a $Y_{\omega,s,J}|Y_{\eta,J}$ and this $Y_{\omega,s,J}|Y_{\eta,J}$ becomes the next $J_{0,J+1}$. So here in [3.2] the bounds are not as in [3.4] but are limits on $Y_{0,J} = f(Y_{\eta,J+1})$.

Out of, and Back into, Chaos

There is a body of strong evidence, mainly due to Freeman and coworkers (reviewed by Skara and Freeman, 1988, see also Globus, 1992) that at least in olfaction the resting state dynamics of the brain in its monitoring of sensory systems are chaotic, and that sensory input pulls down the D_2 Renyi dimensionality to a lower value, by transiently replacing chaos with limit cycling. It is therefore of interest to see if the baseline values a_s, e_s, Y_s can be set to generate chaotic Y trajectories, and move into point attractor dynamics for the induced Γ trajectory, and then restabilize back into chaos. This problem puts a counterintuitive connotation onto 'restabilize', because chaos in the na"ıve sense implies instability, but chaos in its technical sense is a form of stable dynamics; a system sits inside its attractor D_2-bounded space.

There is another small change in the Γ model necessitated here; previously for the Figures 3.1 to 3.3 the value of e was fixed throughout all the η phases at 0.20, for illustration. This value of e is about the middle of the range of values explored in Gregson (1988); that value is associated with ogival psychophysical curves when Y_η is plotted against a. To create chaos for low a the product ae must be raised in value; the e value needed will be normally outside the range used for representing stable trajectories induced by stimulus inputs. So, now two e values are needed, e_s for the baseline state, and e for the destabilizing inputs, with $e_s > e$. In modelling collections of data it would be economical to have one e_s value, but allow a range of e values associated with different stimulus conditions. This point is in fact anticipated in case studies in Gregson (1992a).

However, it is not now possible in this new dynamic context to solve for an a_s that yields a single Y_s, because the chaotic baseline dynamics no longer run onto a point attractor. There is no longer a single Y_s for a given $\{a_s, e_s, \eta\}$ triple; the process is jumping around quasiperiodically on its attractor. The chaotic dynamics are however bounded, the sucessive Y values lie within a range (< 1) and the bounds will envelope any Y_0 or Y_η values generated from stimulus inputs. This holds in any long-term stable situation even if the Γ trajectory of the input itself runs past the first bifurcation. This is easy to represent pictorially, as in Figure 3.4.

The trajectories in restabilization appear to be returning to period 2, but are in fact at least quasiperiod 4 in the simulations pictured.

Refractory Period Predictions

It is possible to interpret the period in which the system returns to baseline, through the action of a_s, as a period within which no new inputs can access the trajectory, provided that the internal variable Y is still above or away from baseline. If it is below baseline then the system might be more readily overwritten with a new input. If this interpretation is accepted, then the minimum observable inter-response interval (i.r.i.) is a function of the duration of two trajectories, the first that runs to Y_η and the second that of the restabilization trajectory. As this sum of two durations is in j (arbitrary) units it can be compared to real empirical data only in terms of the relative i.r.i. values for given stimulus levels. Again, this argument is not readily extended to cope with a chaotic environment; the state after restabilization differs only from its

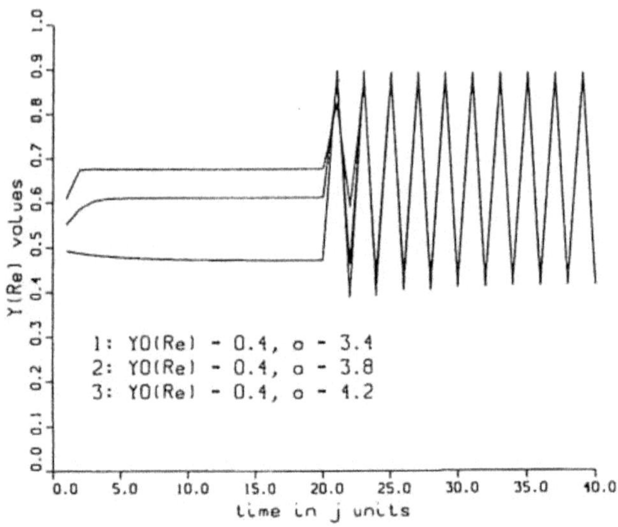

Figure 3.4: Restabilization near chaos

start at $j = \eta$ in the reduction of amplitude of the oscillations in $Y(\text{Re})$.

There is an extensive and somewhat inconclusive argument in the psychophysical literature as to whether refractory periods exist (reviews are by Bertelson, 1966, and Koster and van Schuur, 1973). It is known that stimuli which are intense induce shorter response latencies; here inputs which are far from baseline induce a more rapid arrival at asymptote (if they are not past the first bifurcation) and a quicker return under restabilization. However the present treatment is not extended to choice reaction times, nor to the coupling of motor response to input choice and so the argument from analogy cannot be pursued in detail.

In the present approach, the notion of a refractory period can be given an interpretation in the simpler (non-chaotic) case, but is difficult to substantiate if we incorporate the e_s, e distinction. The extension of these analyses to multidimensional inputs by using $n\Gamma$ theory is possible, but is not pursued here. The question of the stability of the resultant dynamics becomes more difficult to resolve (Silva and Hallam, 1992).

Nonlinear Psychophysics and "Power Law" Scaling

A commonly employed result in psychophysical scaling, nearly always of unidimensional sensory inputs, arises when linear regression plotting of log(response) values on log(stimulus) values, over a restricted range, yields a straight line with second-order disparities, rather like a very flattened ogive. This has been known since the 18th century (Grüsser, 1993) and was rediscovered by the Belgian psychophysicist Plateau around 1870, and repopularised by S. S. Stevens in the 1930's onwards. For this reason it is mistakenly called by some USA writers 'Stevens' Law'.

It is therefore necessary to see if the relationship

$$log(Response) = a + b.log(stimulus) \qquad [3.5]$$

where a and b are constants and b (the exponent) is supposed to characterise a sensory dimension, holds approximately, with second-order deviations as observed, if Y(Re) is substituted for Response in [3.5], and similarly a for stimulus. In fact b is not a constant but has a distribution over subjects and over conditions within experiments. It is easy to show that the log-log Γ $(a \mapsto Y)$ mapping holds for unidimensional stimuli in a middle range; Gregson (1988) gives plots which will satisfy [3.5]. However, a more interesting case is that of judgments of area, where Scharf (1975) and Baird and Noma (1978) estimated that [3.5] holds for judgments of area with an exponent b around 0.7 to 0.8, whereas the exponent for line length (judgments of the length of sides of a rectangle) is around 1.0.

To see if any outputs of 2Γ **Case 2** are compatible with [3.5] when log transformations are imposed on $a_1.a_2$ and on $Y_1.Y_2$(Re) for various λ. Results of doing this for a range of $a_1 \neq a_2$ are shown in Figures 3.5 through 3.7.

The points are produced by 100 combinations of a_1 and a_2, and the lines by varying a_2 for 10 fixed values of a_1. It is seen that linear regressions are generated when λ is near to unity; this is what would be predicted from experimental results on fitting nonlinear theory to other experiments previously reported. The $\{a, e, \lambda\}$ parameter window within which the analogue of [3.5] holds is fairly narrow, and a function of both the component inputs and the cross-coupling. Convex, concave, flatish ogival, and straight lines can arise, and the breakdown when inputs are high is shown in the graph (Figure 3.7) with highest λ.

A metatheoretical comment is pertinent here; the psychophysical 'power law' approximation is not as used a process model, but rather an outcome

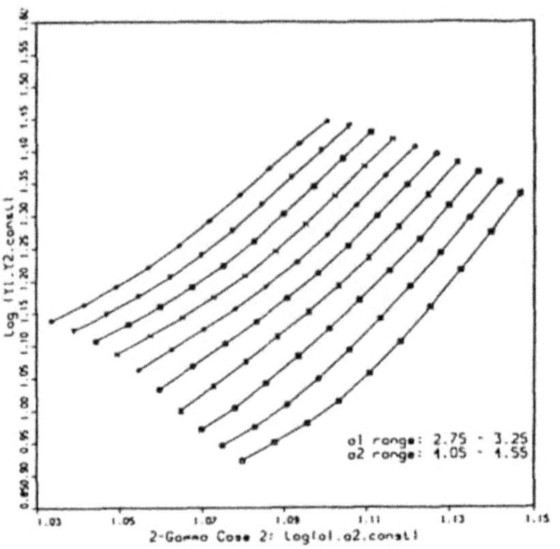

Figure 3.5 Log-Log Plots for 2-Gamma: lambda = 0.85

model of the stimulus-response mappings under stationarity. The Γ trajectory is however a process model, that does not parametrise the externally observable psychophysics, but from which the $a \mapsto Y$ relations can be deduced for a given parameter set, but are not invariant in the simple form that the [3.5] supposes to be general. The invariance in nonlinear psychophysics is in the dynamics of the process, via mappings that can be generated by attractors with fractal geomentry under some parts of the phase space, but it is not meaningful to seek global isomorphisms at the level of traditional psychophysical scalings via Fechnerian (or later) equations. In this sense, the psychophysical dualism of Fechnerian theory, and the dualities of metric space scaling between physical metrics and psychological metrics do not hold, though the algebra of nonlinear psychophysics' Y space can be locally, and only locally, compatible with metric space in the close neighbourhood of the systems' set point.

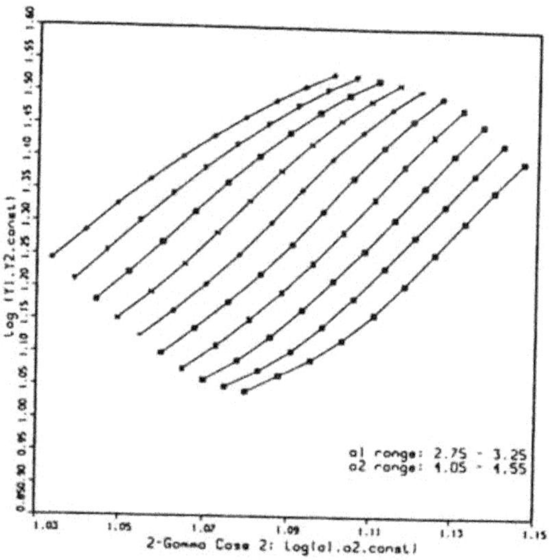

Figure 3.6 Log-Log Plots for 2-Gamma: lambda = 0.95

Dimensional Separability

The idea of dimensional separability[2] in multidimensional perception appears in recent times to have been reintroduced by Garner (1974). The idea of a similarity space representation is a sort of field theory, sometimes assumed to be a metric space, but not the same sort of field as treated in chapters 5 and 6 later. Similarity contour maps are generated in chapter 7. Whether or not dimensions functioning in parallel are separable or can be separable is a theoretical issue that in fact goes back to the dispute between Helmholtz in 1860 and Hering in 1905, already noted in chapter 1, concerning the need to postulate unconscious inferences in perception. Hering and later Wertheimer in 1912 postulated cross-functions (Querfunktionen) which qualitatively anticipate what λ is doing in $n\Gamma$ models. The issue is confounded in experimentation with the choice of response measures used to detect if observers are responding

[2] Some of the work in this section has been previously reported in the electronic journal *Complexity International* in 1994, edited from the Australian National University

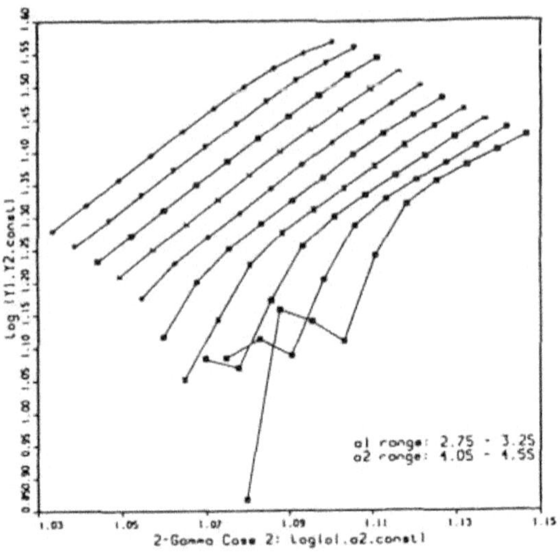

Figure 3.7 of Log-Log Plots for 2-Gamma: lambda = 1.05

to changes in single dimensions within complex stimulation.

Two recent studies[3] (Kadlec and Townsend, 1992, Melara and Day, 1992) illustrate this difficulty, and are interesting for their quite different and contrastable attempts to link the dimensional separability issue into either metric space theories (Melara and Day) or signal detection theory (Kadlec and Townsend). In both cases this enables the authors to draw on an extensive body of measurement theory, if the analysis is valid. In such theory, there are implicit or explicit strong assumptions, and in contradistinction it is better to proceed without making any of these assumptions, if they can be done without. Obviously here we wish to investigate the separability question using nonlinear psychophysical theory and avoid

(a) generating representations of perceptual processes that must be in metric (Minkowski) spaces,

[3] Another version of the Kadlec and Townsend paper is presented in J. Math. Psych., 1992, *36*, 325-374, with the title *Implications of Marginal and Conditional Detection Parameters for the Separability and Independence of Perceptual Dimensions*. The argument is essentially the same.

(b) using Gaussian noise distributions,

(c) restriction to the specific case of two dimensional stimulation,

(d) using stimulus sets with orthogonal properties,

(e) assuming some mappings betwen d' in signal detection and response latencies,

(f) restricting responses to *same-different* dichotomies or to class membership or non-membership attributions.

All of (a) through (f) are in one or other of the analyses offered in the two studies cited.

Melara and Day (1992) focus on the property, in a metric (perceptual) space representation, that for $p = 2$, the Euclidean case, dimensional differences between stimuli are invariant under rotation of the reference axes of the metric space. But in $p = 1$, the city-block space, there is one (and only its $\pi/2$ group transformations) basic orientation of the differences, when they are set parallel to the reference axes themselves.

Let, for example, the stimulus differences, $\Delta U_{2,1} = (U_{11} - U_{12})$, $\Delta U_{2,2} = (U_{21} - U_{22})$ in two-dimensional space be set parallel to the axes x, y where each U_{xy} is defined in terms of the axes. It is assumed that perceived differences corresponding to these ΔU will be monotonic on d' if we take an SDT approach (Kadlec and Townsend) and will be inversely monotonic on response latencies for discrimination (Melara and Day) in classification tasks where the stimulus set is divided into two groups of two, and various rules dependent sufficiently on the composition of the two-component stimuli are used to define the classification tasks.

To test the idea that if stimuli are in Euclidean space then classification is more efficient, Melara and Day used sets of four stimuli with vibrotactile *perceptual* dimensions of pitch and loudness, and generated some by rotation of the orthogonal configuration through angles of $\pi/8$ and $\pi/4$, thus not exploring the periodicity of the $\pi/2$ rotational group transformation.

The rotation is effected by the matrix multiplication

$$\begin{pmatrix} U'_{21} & U'_{22} \\ U'_{11} & U'_{12} \end{pmatrix} = \begin{pmatrix} cos\theta & sin\theta \\ -sin\theta & cos\theta \end{pmatrix} \begin{pmatrix} U_{21} & U_{22} \\ U_{11} & U_{12} \end{pmatrix} \qquad [3.6]$$

Such rotations can only be performed in spaces with even dimensionality, because a rotation in a space of an odd number of dimensions always leaves a one-dimensional subspace invariant (Murdoch, 1957). Hence the argument does not readily generalize to stimulus comparisons of higher dimensionality; three dimensions would be a more critical test. The error rates on the tasks were very low, the general conclusion (p. 14) is that 'perceivers process multidimensional stimuli through a set of primary dimensions, even when the stimuli are formed from dimensions that interact'. The orientation of the stimuli did not change the role of primary dimensions, in this two-dimensional case.

The Kadlec and Townsend study also reviews a number of experiments. It essentially advances a series of conditional definitions of comparisons generating d' measures in an SDT framework imposed on pairwise comparisons of orthogonal Gaussian p.d.f.s. Each comparison may be conditioned on a fixed level of the dimension orthogonal to it. In short, each U_{hk} is replaced by a bivariate p.d.f centred on $\bar{U}_{hk}$, and the marginals on this bivariate p.d.f. are then used for conditional distributions. Kadlec and Townsend adopt the strong Gaussian assumption in their treatment, but observe that it is not a necessary parametrization.

The separability condition in the SDT analysis is replaced by marginal response invariance, and useful distinctions are drawn between perceptual separability and decisional separability. The analysis extends to truth value tables for separability and perceptual independence, and is applied to cases where stimuli have two, and four, components, drawing on data of Townsend, Hu and Evans (1984). and Nosofsky (1985,1986). Not all the results are presented in a manner which facilitates re-analysis without SDT assumptions. The SDT approach yields post hoc tests to compare conditions; it does not predict *a priori* d' values, but rather predicts the pattern and shifts in d' values for different elicited judgments.

Four orthogonal U stimuli, arranged as in [3.6], have one-dimensional separations horizontally or vertically (that is, parallel to the reference axes as in [3.6,3.13]); the only way to introduce comparisons involving differences on both component dimensions is to consider the diagonal contrasts $U_{11,22}$ and $U_{12,21}$.

If the configuration is rotated in the a plane, the inter-U distances being maintained (that is, a transformation under rotation without dilation, nor with movement of the centroid $\bar{U}$ relative to the axes origin) then all six pairwise U comparisons become two-dimensional. After rotating through $\pi/4$ two comparisons become one-dimensional, that is, the diagonal contrasts. This rotation induces a group G_θ with periodicity $\pi/2$, jumping between 4 unidimensional and 2 bidimensional contrasts, 4 bidimensional, 2 unidimensional and 4 bidimensional, 4 bidimensional, and so on. The effect, in stretching or shrinking the $U_{vv',ww'}$ distances in a-space, differs according to whether the a-space is

City-Block ($p = 1$) or Euclidean ($p = 2$); only the latter having distances invariant under rotation. One can intuitively see why this is so, for the isodistance contours for $p = 2$ are circles, for other $p = 1, \infty$ they have singularities. It is this property which Melara and Day utilise.

As noted previously, this G_θ is associated with even dimensionality of the U. Also, in non-Euclidean space, the group transformation G_θ for dimensionality of contrasts is coupled with another group transformation of comparable periodicity for interstimulus distances. If $n\Gamma$ is coupled with a model of similarity (Gregson, 1975,1993c,1994) the exact predictions of pairwise similarity values between n-dimensional stimuli can be derived, for given stimulus compositions. The $n\Gamma$ model [1.5-1.7] specifically allows for interactions between input channels in parallel; whether or not such interactions have any detectable consequences depends, in that model, on whether the cross-coupling between dimensions is compensated (Gregson, 1992a, p. 25). If it is compensated then the system behaves as though the channels are independent, but this is a special case not readily generated exactly.

A Generalized Nonlinear Model

A set of stimuli are created, each with specifiable components $\{a\}$, one of which is fixed at the same value for all stimuli, and their comparisons treated as generated by a 3Γ process. Let there be four stimuli, $\{U_{11}, U_{12}, U_{21}, U_{22}\}$ and let each stimulus have three components (all stimuli are composed on the same three sensory dimensions). So one stimulus, U_{hkw} is defined as

$$U_{hkw} = \{a_h, a_k, a_w\}$$

with $a_w = a_0$ =constant and second-order. Because the third component is fixed, the third dimension is a point and in variability terms the system becomes 2Γ but with an added low fixed component $a_w \rightarrow 2.001$, in a 3-dimensional $\{a\}$ context. The cross-coupling λ will modify outputs between all three a. We have called this situation $2-(3)\Gamma$.

Rotational transformations can be performed on an $n-(n+1)\Gamma$ configuration, if n is even, because the odd extra dimension is in a point subspace which can only be reflected.

In Table 3.1 the effect of the δa is to induce rotation only if their values are set in a particular pattern. Rotation through $\pi/4$ is simplest, and is used later in Table 3.4. This use of δa values is unrestricted in terms of orthogonality and rotation, and is thus more general than the matrix operation in [3.6].

Table 3.1:
Coordinates of Four Stimuli on the First Two Dimensions

Each component $a_{xy,z}$ has dimension z and U-coordinates x, y. The differences $\delta_{\xi,z} \geq 0$ have type ξ and dimension z; $z \in \{g, h, w\}$. Type is a dummy variable. From the constraint of $2-(3)\Gamma$, all $a_{xy,w} = a_0$.

$$\begin{array}{lcccc} a_{12,1} & = & a_{11,1} & + & \delta_{1,1}a \\ a_{22,1} & = & a_{21,1} & + & \delta_{1,1}a \\ a_{11,2} & = & a_{12,2} & + & \delta_{1,2}a \\ a_{21,2} & = & a_{22,2} & + & \delta_{1,2}a \\ a_{11,1} & = & a_{21,1} & + & \delta_{2,1}a \\ a_{12,1} & = & a_{22,1} & + & \delta_{2,1}a \\ a_{21,2} & = & a_{11,2} & + & \delta_{2,2}a \\ a_{22,2} & = & a_{12,2} & + & \delta_{2,2}a \end{array}$$

Some further definitions are required:
In $2-(3)\Gamma$, let

$$\psi(\Delta(U, U') = f(Y_h - Y_{h'}, Y_k - Y_{k'}) \qquad [3.7]$$

Let a similarity $\mathcal{S}$, $0 \leq \mathcal{S} \leq 1$, be, for any two stimuli V, W, and $i = 1, 2, .., n$ dimensions with components u

$$\mathcal{S}(V, W) = \frac{\sum_{i=1}^{n} \alpha_i \cdot min(u_{vi}, u_{wi})}{\sum_{i=1}^{n} \alpha_i \cdot max(u_{vi}, u_{wi})} \qquad [3.8]$$

with $\sum \alpha_i = 1$. This is a set theoretic model (Gregson, 1975, eqn. 5.44.1), of a dimensionless similarity number, it assumes that each physical attribute dimension is known and can be measured, and that the psychological information being used by the observer is covered by those measures. In the sense of Melara and Day (1992) V and W are *primary dimensions*. To be precise $\mathbf{m}(u_{vi})$ should be written for u_{vi} (and so for all v and w terms in [3.7]), distinguishing the measure $\mathbf{m}$ from the attribute u in each case. The isosimilarity contours for [3.8] are neither squares nor circles, but are trapezoids; they are confusable with $p = 1$ isodistance contours of the Minkowski spaces. For [3.8], there are four linear segments of the isosimilarity contour, with four singularities. The $n\Gamma$ and $\mathcal{S}(V, W)$ processes are put into cascade, with $n\Gamma$ coming first. The u

values in [3.8] are thus replaced by their corresponding $Y(\mathrm{Re})$ values from the output of an $n\Gamma$ process, and the resultant $\mathcal{S}_\Gamma$ rather than [3.8] are used in the simulations here. The shape of the contours for $\mathcal{S}_\Gamma$ has been derived, and is mixed, concave, convex, and a labile function of η, λ, and $min(a)$ (Gregson, 1993c). It is also necessary to have a definition of relative similarity, if comparable class membership choices are to be elicited. This we can take from Gregson (1975) as $\mathcal{R}$ at L^2, which is

$$\mathcal{R}_{w,z} = \mathcal{S}_{v,w}/\mathcal{S}_{v,z} \qquad [3.9]$$

Note that $0 \leq \mathcal{R} \leq \infty$; it is not bounded like $\mathcal{S}$. Similarity varies in degree; it is only in the degenerate case dichotomous. Researchers who dichotomise response decisions, as in Kadlec and Townsend (1992) therefore offer data which are of very limited use, unless an ancillary assumption is made that

$$\mathcal{S} = f(d') = exp(c \cdot -d') \qquad [3.10]$$

though no precise data on this mapping appears to exist. It may be noted that both sides of [3.10] are dimensionless, the elimination of physical dimensions implicit in a distance is achieved by using d' as a standardised difference.

Table 3.2:
First-Order Pairwise Similarities and Second-Order Relative Similarities
$\mathcal{S}_{vv',ww'}$ and $\mathcal{R}_{vv',ww'|zz'}$

$\mathcal{S}_{11,12}$		
$\mathcal{S}_{11,21}$	$\mathcal{S}_{12,21}$	
$\mathcal{S}_{11,22}$	$\mathcal{S}_{12,22}$	$\mathcal{S}_{21,22}$

$\mathcal{R}_{11,12\|21}$			$\mathcal{R}_{11,12\|22}$	-	-
$\mathcal{R}_{11,21\|12}$	$\mathcal{R}_{12,21\|11}$	-	$\mathcal{R}_{11,21\|22}$	$\mathcal{R}_{12,21\|22}$	
$\mathcal{R}_{11,22\|12}$	$\mathcal{R}_{12,22\|21}$	$\mathcal{R}_{21,22\|11}$	$\mathcal{R}_{11,22\|21}$	$\mathcal{R}_{12,22\|11}$	$\mathcal{R}_{21,22\|12}$

If some secondary blurring of the U coordinates is admitted, which in effect is what Kadlec and Townsend do in invoking Gaussian distributions centred at each U_{hk}, then correspondingly we would wish to introduce imprecision on the rotation parameter, θ in [3.6] which underlies the group transformation, so that the probability of landing exactly on a θ value of $0, \pi/4, \pi/2, 3\pi/4, \ldots$ is asymptotic to zero. The situation then stochastically reduces to one in which all stimulus comparisons in n-space are n-dimensional, not $n-1$.

This point matters, because Kadlec and Townsend consider conditional marginal d' distributions, with $n = 2$, for one-dimensional comparisons given that the other dimension a value is fixed high or low as the case may be. If θ is not fixed, then no a value is fixed either, and the contextual effect of a supposedly fixed dimension (like the actually fixed third dimension we are using here in $2-(3)\Gamma$) is better described as that of a second-order difference; the first-order differences for d' are then conditional on a second-order d'. Actually the Gaussian distributions of Kadlec and Townsend can be replaced by a non-Gaussian n-dimensional p.d.f induced by instabilities in θ.

The argument for conditional separability, when in n dimensions $n-1$ are fixed as point constants a_c, in a given m-way comparison, is not quite the same as the $p = 1$ and $p = 2$ contrast, but is related to it. Where $\mathcal{D}$ means *discriminate in degree*, consider the cases

$$U(a_{11}, a_{21}) \; \mathcal{D} \; U(a_{11}, a_{22}) \qquad [3.11]$$

$$U(a_{11}, a_{21}) \; \mathcal{D} \; U(a_{12}, a_{22}) \qquad [3.12]$$

If conditional separability exists, then:

The basis for overall $\mathcal{D}$ for [3.11] is $f(a_{22} - a_{21} | a_{11})$.

The basis for overall $\mathcal{D}$ for [3.12] is $f(a_{22} - a_{21}, a_{12} - a_{11})$.

The basis for conditional $\mathcal{D}$ for [3.11] is $f(a_{22} - a_{21})$.

The basis for conditional $\mathcal{D}$ for [3.12] is $f(a_{22} - a_{21} | a_{12} - a_{11})$.

As the variance of a_{11} approaches $a_{12} - a_{11}$ the overall $\mathcal{D}$ for [3.11] and [3.12] become indistinguishable. If dimensions are not separable then the same holds for the conditional $\mathcal{D}$.

Simulations

The independent variables to explore are $\delta_{1,1}, \delta_{1,2}, \delta_{2,1}, \delta_{2,2}$ and λ, given a set of a inputs. The dependent variables are $\mathcal{S}$ and $\mathcal{R}$ over all the cells in Table 2. The a_{xyz} are constants throughout.

A: In the case $\forall \xi,\ \delta_{\xi,z} = 0$, we have what corresponds to the orthogonality condition, the dichotomous separability is replaced by a variation in degree of the cross-coupling λ; increasing λ would correspond to increasing imperfect separability.

B: In the cases where $\delta_{\xi,z}$, $z = 1,2$ is replaced by the matrix operation in [3.6] we can have rotation which yields something analogous to Melara and Day's work.

C: In the case

$$\delta_{\xi,z} \geq 0, |\delta_{1,z}| \neq |\delta_{2,z}|$$

we can have dilation (or contraction) and some rotation. **C** is general, **A** and **B** are special cases of **C**.

We therefore consider the last case.

Table 3.3 is an introductory example, to illustrate the patterns of all the $\mathcal{S}_\Gamma$ similarities obtainable from four othogonal $2-(3)\Gamma$ stimuli compared using [3.8]. There is no rotational transformation, the variables of interest are η and λ. It is necessary to consider η (the number of iterations in the trajectory) because for high a values the Γ trajectory goes into limit cycling, and odd and even η induce different Y outputs. The effect of λ is to vary separability. The four stimuli, in the 2×2 array, with a components,

$$\begin{pmatrix} a_{22} & | & U_{21} & U_{22} \\ a_{21} & | & U_{11} & U_{12} \\ ---- & & --- & --- \\ a_{2\bullet} \Uparrow a_{1\bullet} \Rightarrow & | & a_{11} & a_{12} \end{pmatrix} \qquad [3.13]$$

yield the comparisons $U_{vv',ww'}$ as in Table 2 for $\mathcal{S}$.

Low λ values correspond to weak separability, and
High λ values correspond to strong separability.

There is no complete fusion at one extreme, nor complete separation at the other. Pushing λ beyond the limits used here leads to explosion in the Γ recursions. Only three low λ and three high λ settings are shown in Table 3.3, for constrast. In between the cell ($\mathcal{S}$) values are monotone increasing or decreasing with λ

Graphically the a coordinates used were

3, 4.6	4, 4.6
3, 3.6	4, 3.6

The diagonal comparisons U_{11}, U_{22} and U_{12}, U_{21} are labelled as such in Table 3.3.

The differences between corresponding $\mathcal{S}$ values for low λ and high λ (the relative effect of degrees of separability) is shown in Figures 3.8 through 3.11 for

Table 3.3:
Examples of Similarities with odd and even iterations
a values 3.000 and 3.600 δa values 1.000 and 1.000
2-(3)Γ **Case 2**, $\eta = 10$ on left and $\eta = 11$ on right

		$\lambda = 0.75$			
0.483			0.497		
0.682	0.512diag		0.676	0.520diag	
0.589diag	0.575	0.812	0.584diag	0.583	0.813
		$\lambda = 0.80$			
0.507			0.509		
0.729	0.537diag		0.724	0.537diag	
0.633diag	0.594	0.822	0.631diag	0.593	0.825
		$\lambda = 0.85$			
0.575			0.496		
0.771	0.587diag		0.769	0.538diag	
0.674diag	0.593	0.829	0.677diag	0.589	0.835
		$\lambda = 1.20$			
0.525			0.747		
0.884	0.531diag		0.876	0.777diag	
0.844diag	0.581	0.921	0.739diag	0.735	0.792
		$\lambda = 1.25$			
0.586			0.472		
0.901	0.693diag		0.871	0.490diag	
0.857diag	0.706	0.906	0.751diag	0.408	0.789
		$\lambda = 1.30$			
0.670			0.668		
0.901	0.652diag		0.858	0.725diag	
0.874diag	0.778	0.899	0.770diag	0.776	0.786

$\eta = 10$ and 11. The variability between odd and even η substitution means that, to an external observer, the predicted variance of some $\mathcal{S}_\Gamma$ values will be much greater than for others. Hence, if a simulation over an η, λ parameter space is generated, then $var(\mathcal{S})$ values can be derived. We postpone this question in deference to the separability and orthogonality problems.

It appears that in the case selected for Table 3.3, diagonal $\mathcal{S}_\Gamma$ comparisons (that is, differences generated on two a components and not just on one) are

Figure 3.8:
Similarities Represented in Coordinate Space.
2-(3)Γ Case 2, $\eta = 10$, $\lambda = .75$
Low Separability Case as in Table 3.3

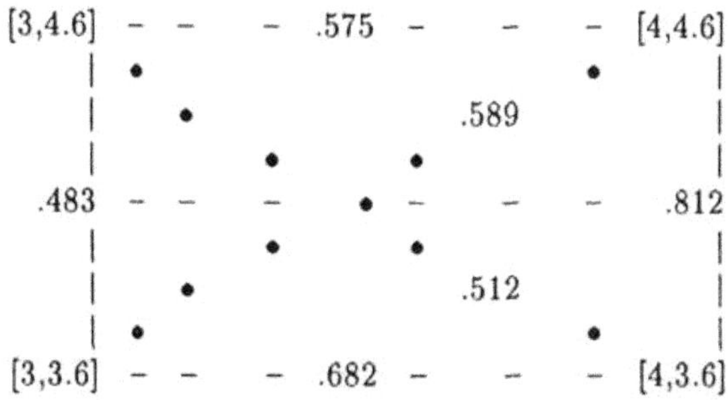

Figure 3.9:
Similarities Represented in Coordinate Space.
2-(3)Γ Case 2, $\eta = 10$, $\lambda = 1.30$
High Separability Case as in Table 3.3

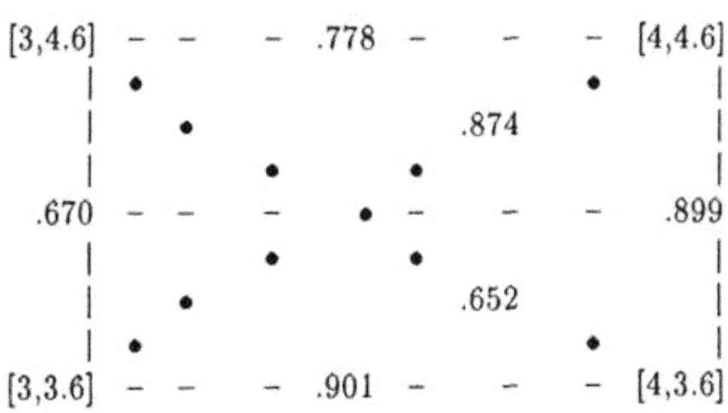

Figure 3.10:
Similarities Represented in Coordinate Space.
2-(3)Γ Case 2, $\eta = 11$, $\lambda = .75$
Low Separability Case as in Table 3.3

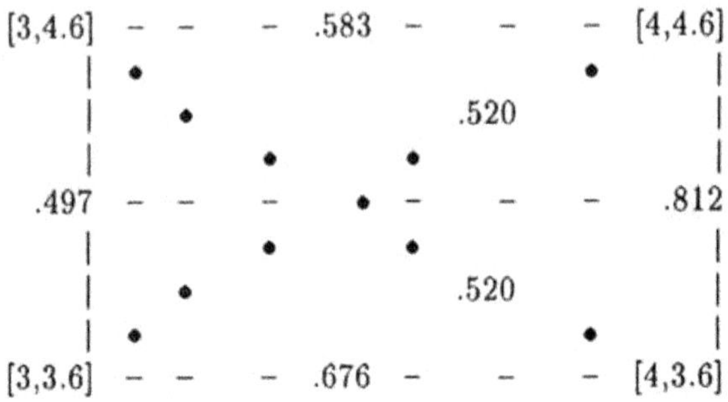

Figure 3.11:
Similarities Represented in Coordinate Space.
2-(3)Γ Case 2, $\eta = 11$, $\lambda = 1.30$
High Separability Case as in Table 3.3

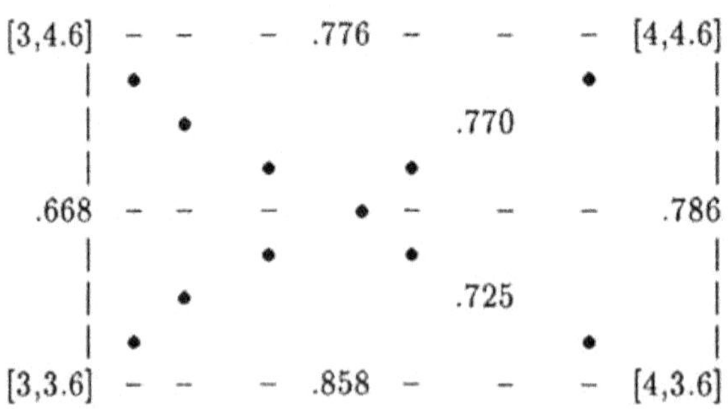

more susceptible to separability (λ) effects than the comparisons parallel to a principal axis. This is hardly surprising, but it is an empirically testable prediction. It suggests that the discussion by Melara and Day may be confusing two effects, one related to the number of unidimensional differences involved in a multidimensional stimulus comparison, and the other to the Minkowski metric's properties. This point we anticipated in introducing the group G_θ.

The effects of λ are not independent of the stimulus compositions used to create the contrasts and similarities. A series of figures parallelling Figures 3.8 through 3.11 is readily created using, instead, $a_{11} = 3.0, 3.8, a_{12} = 2.4, 3.8, a_{21} = 3.0, 3.6, a_{22} = 2.4, 3.6$. We call this case 'new coordinates', Figures 3.12 through 3.15 correspond respectively the parameter settings for Figures 3.8 through 3.11. The λ range is set slightly lower to avoid explosions in Y trajectories. Again, $\eta = 10$, or 11, are employed for even and odd $2-(3)\Gamma$ iterations.

In this second example, where the a values are set lower in one direction, the effects of η are second-order for high separability (compare Figures 3.13 and 3.15).

Table 3.4 illustrates what happens if the stimulus coordinates are rotated through $\pi/4$. The meaning of the δa terms is now slightly different from that in Tables 3.3 and 3.4. We have shown more λ steps in Table 3.4 than in Table 3.3, because the separation of the values for η, for some $\mathcal{S}$, occurs at lower λ than is the case for Table 3.3.

Now the four sides of the square on U become diagonals with respect to the axes,and the diagonals become parallel to the coordinate axes. Again, the diagonals are longer than the sides, but now the diagonals involve unidimensional comparisons, and the oblique sides involve two component differences. The linkage between distance and dimensionality within the set of six comparisons is reversed. The more marked effects of λ are now not shown on the diagonal similarities. Figures 3.16 and 3.17 show the λ extremes pictorially for $\eta = 10$.

The effects of λ are (as in Figure 3.16) quite dramatic; a shift of $\mathcal{S}_\Gamma$ from .455 to .963, for the $U[3.0, 3.8 - 3.707, 4.507]$ comparison is interestingly in contrast to the near stability of the shift from .868 to .890 for the opposite side of the lozenge.

Table 3.4:
Examples of Similarities with odd and even iterations
The rotated case: clockwise turn through $\pi/4$
a values 3.000 3.800 δa values 0.707 0.707
2-(3)Γ **Case 2**, $\eta = 10$ and $\eta = 11$

			$\lambda = 0.70$		
0.455			0.444		
0.560	0.520diag		0.557	0.514diag	
0.604diag	0.679	0.868	0.600diag	0.670	0.869
			$\lambda = 0.75$		
0.534			0.521		
0.592	0.572diag		0.588	0.565diag	
0.635diag	0.725	0.880	0.630diag	0.716	0.881
			$\lambda = 0.80$		
0.613			0.603		
0.630	0.623diag		0.627	0.620diag	
0.674diag	0.772	0.889	0.668diag	0.767	0.890
			$\lambda = 0.85$		
0.683			0.677		
0.670	0.665diag		0.670	0.667diag	
0.715diag	0.812	0.895	0.710diag	0.810	0.898
			$\lambda = 1.10$		
0.854			0.849		
0.833	0.826diag		0.752	0.722diag	
0.834diag	0.889	0.962	0.829diag	0.889	0.888
			$\lambda = 1.15$		
0.878			0.853		
0.844	0.867diag		0.753	0.714diag	
0.852diag	0.893	0.959	0.830diag	0.893	0.879
			$\lambda = 1.20$		
0.916			0.845		
0.875	0.894diag		0.773	0.715diag	
0.875diag	0.863	0.938	0.819diag	0.897	0.878
			$\lambda = 1.25$		
0.963			0.839		
0.915	0.909diag		0.790	0.716diag	
0.842diag	0.854	0.890	0.811diag	0.901	0.877

Figure 3.12:
Similarities Represented in Coordinate Space.
2-(3)Γ Case 2, $\eta = 10$, $\lambda = .70$
Low Separability Case, New Coordinates.

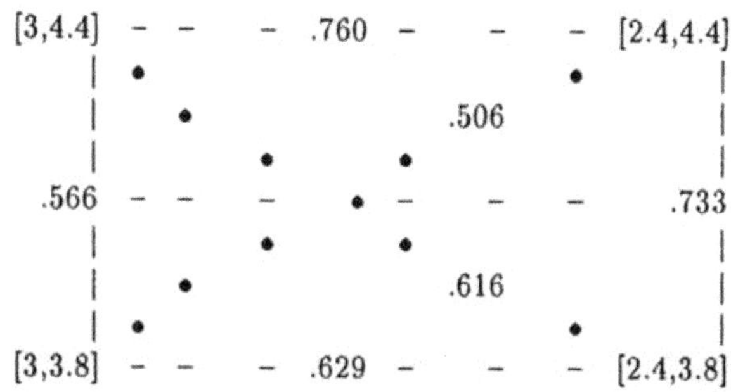

Figure 3.13:
Similarities Represented in Coordinate Space.
2-(3)Γ Case 2, $\eta = 10$, $\lambda = 1.25$
High Separability Case, New Coordinates.

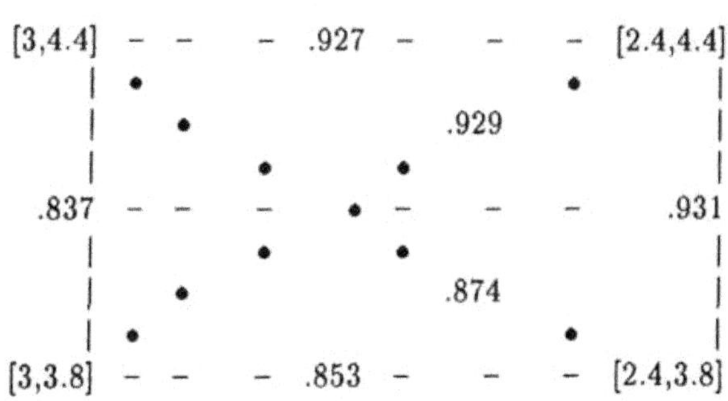

Figure 3.14:
Similarities Represented in Coordinate Space.
2-(3)Γ Case 2, $\eta = 11$, $\lambda = .70$
Low Separability Case, New Coordinates.

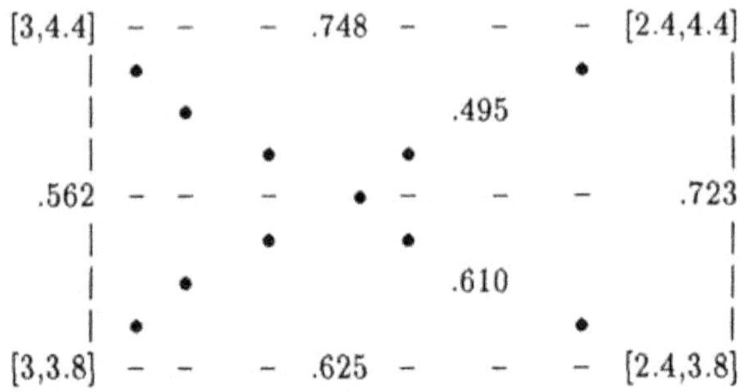

Figure 3.15:
Similarities Represented in Coordinate Space.
2-(3)Γ Case 2, $\eta = 11$, $\lambda = 1.25$
High Separability Case, New Coordinates.

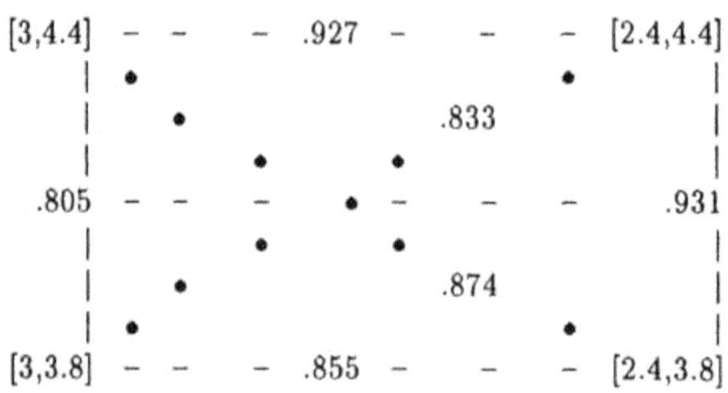

Discussion

From these examples, which are constructed using a theoretical response measure which is not the same as those employed in the Melara and Day, or Kadlec and Townsend studies, but is more sensitive to differences in the information content of the psychophysical system's output, we may deduce that the effects of cross-coupling between stimulus dimensions will modify similarities. These similarities are in turn also functions of

(a) the stimulus dimensionality of overall differences,

(b) the distance in coordinate space of the physical stimulus separations, because this affects substitutions in $\mathcal{S}$ and $\mathcal{S}_\Gamma$,

(c) the actual coordinate values, because the $n\Gamma$ system runs into quasiperiodicity or chaotic dynamics for high inputs,

(d) the parity (odd,even) of the dimensionality if rotation is used in some dimensions.

There is a duality between $p = 1$ and $p = \infty$ in 2-space upon which Melara and Day do not comment; this is in terms of configurations of distance representations in metric space. If $p = 1$ fits data well, then so does the supremum metric $p = \infty$.

Some of the results here suggest that pseudoseparability could be generated for a suitable choice of system parameters. In terms of [1.5-1.7], **Case 2** can mimic **Case 1** for some λ, η. The situation in Figures 3.16 and 3.17, and their corresponding previous figures, even suggests that some pairwise comparisons in a set might look separable, and others in the same set not so. Separability, or its analogue in $n\Gamma$ theory, is not necessarily a global property of the parameter space.

The consequences of this treatment lead to the prediction that the shape of isosimilarity contours about a reference point (a_{r1}, a_{r2}) is not invariant, and is a function of λ. This matter is further explored in Chapter 7.

Figure 3.16:
Similarities Represented in Coordinate Space.
2-(3)Γ **Case 2**, $\eta = 10$, $\lambda = .70$
Low Separability Case, Rotated Coordinates.

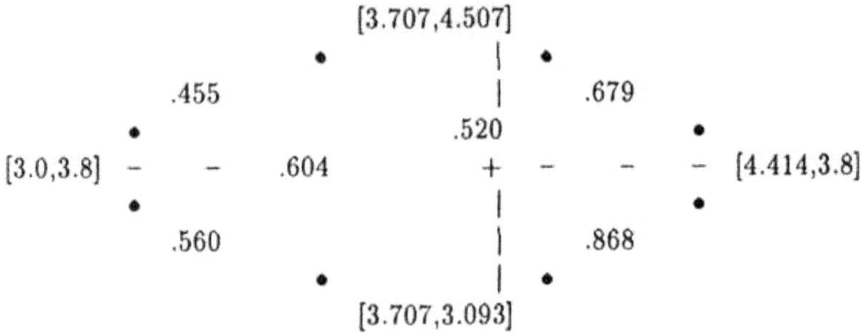

Figure 3.17:
Similarities Represented in Coordinate Space.
2-(3)Γ **Case 2**, $\eta = 10$, $\lambda = 1.25$
High Separability Case, Rotated Coordinates.

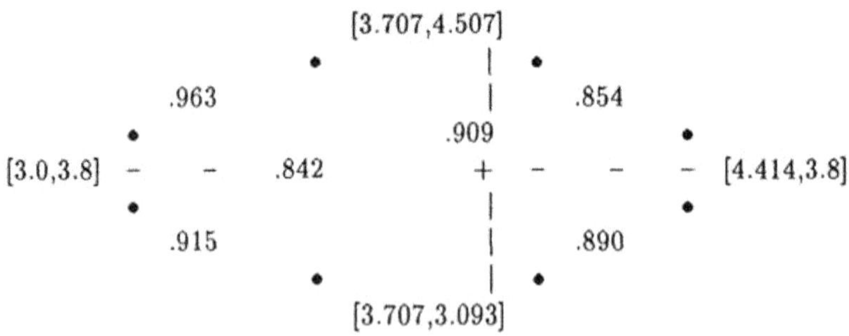

4 Cascades and the case of Perceived Time Intervals

Time like an ever rolling stream
Bears all its sons away,
They fly forgotten as a dream
Dies at the opening day.

From a hymn by Isaac Watts, 1719

Cascades originally meant waterfalls which were broken into a series of smaller waterfalls with intermediate pools. It is not too fanciful to remark on the dynamics of such a cascade, as a preliminary to using variants of the idea in psychophysics. The water stream is continuous, but at intermittent and not necessarily equally-spaced points the kinetic energy is converted into potential energy, or partly dissipated in irreversible local turbulence. The memory of the dynamics is destroyed at each pool, and recreated *ab initio* in each new waterfall. The flow within each small waterfall is then in theory describable by equations of smooth or turbulent flow. In one sense the whole system has continuity, the same water comes out at the bottom as went in at the top. If the input is cut off then there is eventually no output. How many of these features are echoed in a cascade of electrical, chemical or informational measures depends on the assumptions of the cascade as a metaphor. In psychophysics it is expedient to begin again from first principles, and not to borrow intact some algebra from the physical sciences. Not all the details of a physical cascade, as a metaphor, necessarily carry over into psychological theorising.

Ubiquitous Time

The idea of time features in experimental psychology in different senses, as a central variable in processes, as merely a counter-variable for tracking processes, and as a phenomenological experience which is the intended psychophysical focus of experimental studies. These can be brought together within one case study in this chapter.

One of the ideas to which theoreticians in mathematical psychology come back again and again, in one form or another, is that of hierarchical dominance of subprocesses by a higher-level process. By hierarchy is meant that there are at least two levels, one of elementary processes and another which has its own dynamics and acts as a controller, switching the elementary processes in and out and selecting between them. Such selection can arise externally as a result of feedback from the consequences of previous system outputs. Note that this is not the argument about the minimum number of levels in a distributed network which are necessary and sufficient for learning, but to something even more abstract. The analogy with the structure of massive parallel computers may be noted, but it is not a tight analogy.

Traditional psychophysical models such as Fechner's, or later Plateau's in the 1870s, are by definition outcome models. They rest on the only statistics available to Fechner in the 1850s, partitioning a linear function into trend and residual gaussian error. The additional assumptions that are needed to justify this seem to be buried under a conspiracy of silence between editors and authors, so let us remind ourselves that stationarity of the input-output mappings generated in time is a condition of pooling data over time, and the nonstationarity of such psychophysical effects as contrast and assimilation have led to attempts to treat such apparent local nonstationarities as the consequence of the wide sense stationarity of autoregressive processes underlying series of responses. We are still at the outcome level, but now time has come back in to be used as a counter variable. One gets some fun out of switching in the dualities between between-event series in real time steps, and time-interval series resting on an event counter, but it all gets rather opaque as to what the autoregressive spectra actually mean. Still, the point is worth making that an outcome model which fixes the invariance of trial-to-trial sequential dependencies is itself creating slow dynamics which must be deducible from a process model of the relatively fast within-trial dynamics. Again, time series models are about observable outputs on a trial-to-trial basis but look deceptively like process models because time orders have crept back in. Actually the dynamics get lost in the lags representation of an autoregessive spectrum, or the frequency domain decompositions that are equivalent and simply statistical shorthand, given wide-sense stationarity. It is unfortunate that in many studies Fourier decompositions are advanced as sufficient process descriptions of information transmission, when they are not unless we can show in detail that a network functions exactly as a spectral filter (Dodwell and Caelli, 1993).

If a process model is written for within-trial events it must generate all the second-order input-output non-veridicalities which characterise human sensation and perception, as an interconnected bundle, and not as in outcome models set up a separate equation for each anomaly, such as departures from Weber's Law, thresholds, response latencies, time order errors, psychometric functions, and other phenonomena which complicate matters.

Cascading and Sequential Perception

The basic trajectories of Γ recursions may be used as little building blocks to construct what is really an hierarchical model of sequential perception. The idea of Gamma Zak Embedding (GZE) (Gregson 1993a) which leads to processes with response delay distributions which are not strictly periodic nor necessarily unimodal is implicit here. As [1.1] is a transient destabilisation of Y by modifying locally the gain of the attractor dynamics, modifying the gain alters the qualitative dynamics and hence the phase-space portrait of the system, and when the input is used up the system restabilises; this again is not instantaneous, as discussed in chapter 3. So the first two building blocks are

destabilisation (stimulus input)

restabilisation (return to dynamic baseline)

The Γ trajectories are quite brief, given a point input (a Dirac delta function). For an input which is to be temporally extended and thus sustained in time, it is necessary to use an operation called

superposition (inputs extended in time)

where each re-setting of the gain modifies the trajectory unless the trajectory is at a stable asymptote, and a fourth operation

cascading (autonomous perseveration)

where the output state of the trajectory is fed onto the gain of a following (in time) trajectory with the same parameters. This cascading does not need more than one initial input to get it going. All these four blocks are actually subprocess models; the GZE idea imposes a time base on their evolution. By using terminal attractor theory, something can be thought of as network behaviour but certainly not as single neuron modelling.

Such ideas are obviously not unique to this account; loose analogies can be found in a lot of modern system theory. The specific feature here is that they are all variants of their Γ recursion algebra, which has only two parameters, the gain a which is input dependent, and the second internal parameter e which modifies sensitivity. From any serial combination of these four blocks one could derive an outcome prediction over the total response surface of the process. The traditional psychophysical functions fall out nicely but with second-order disparities which are also observed in real data. Obviously (or it should be obvious) one cannot work backwards from Y_{obs} outcome results to the Γ models and get a unique solution to the internal mapping.

Prospective Time Duration Estimation

The experimental and theoretical literature on the psychophysics of time duration estimation is copious, and goes back at least to Vierordt (1868) and his eponymous law, that short durations are overestimated and long durations are underestimated. Discussions on change, time, the present *now*, and the measurement of time both in the sense of time being measurable and time providing a measure of change, are older, dating from commentaries on Aristotle's *Physics* by the neoplatonists around 500 A.D. These commentaries interestingly anticipate, often in some detail, the arguments at a verbal level of contemporary cognitive psychologists (Urmson, 1992), so that one may be disposed to discount some of the apparent progress in modern discussions of the topic.

Over 1200 publications[1] now exist (Zelkind and Sprug, 1974); thorough reviews by Woodrow (1951), by Allan (1979), and by Hellström (1985) are the appropriate starting points; a little more of substance has since been added (e.g. Curtis and Rule, 1977, Wearden and Ferara, 1993), but so many of the possibilities in choice of experimental designs, and in post hoc explanations, had already been canvassed up to and including the studies of Needham (1934,1935). The focus here is on what happens when two time intervals are presented to an observer, usually in series but sometimes in parallel, and he or she is asked to estimate their relative lengths. The desired response may in its simplest form simply be 'longer', 'shorter', or 'equal', or more precise estimations of the amount of the difference between the two durations and associated confidence limits on that judgment may be elicited. This sort of experiment is called prospective, in that the observer is told in advance that time duration

[1] Prof. Hannes Eisler of Stockholm informs me he has identified over 5000; hopefully for the busy researcher most would not be relevant to this analysis.

estimations are going to be required. The distinction between retrospective estimates, made after an elapsed interval, without prior knowledge that an estimate is going to be required, and prospective estimation, induces quite different psychophysiological functions (Hicks, 1992). The psychological distinction between retrospective and prospective time experiences was originally drawn by James (1890).

Some attempts at linear models where a duration has an internal scalar representation have been offered (Allan and Kristofferson, 1974, Eisler, 1981) but are known to be wrong in some of their predictions. Those scalar models are outside real time, they are purely input-output mappings with no internal dynamics, in contradistinction to trace theories which do postulate an internal evolution of processes. As it is known that the results of prospective experiments are anything but consistent, and are apparently functions of the durations and the temporal interval between stimuli, and of some contextual variables considered to be both perceptual and cognitive (Block, 1990, Macar, Pouthas and Fiedman, 1992), a notation is first set up in Table 4.1 as a framework. What in effect is being done here is to rewrite a type of time estimation model that already exists, in various forms, but now using the building blocks of nonlinear psychophysical theory to represent component stages of the model. So the broad framework of the approach is not completely original, but its particular algebraic expression is new to time studies. Gibbon, Church, and Meck (1984) for example described an internal-clock model as Scalar Timing, and noted that such a model needs three stages; a pacemaker, a switch to move pulses from a pacemaker to an accumulator, and an integrator that stores what comes into the accumulator. This sort of structure is not unique to Scalar Timing, and features in other models[2] (Creelman, 1962, Killeen, 1992).

When marked individual differences are noted in behaviour, as a function of arousal, drug induced states, or similar factors, this is commonly postulated to be due to induced changes in the pacemaker part of the model; the internal clock can be speeded up or slowed down. Initially, therefore, we take the clock as running at a fixed rate, and defer tracing out the implications of accelerating it (positively or negatively) as a second problem contingent upon a solution to the model's structure.

---

[2] Hicks, Miller, Gaes and Bierman (1977, p. 442) write of "a time base, which is responsible for subjective temporal units, a counter which stores these units until the conclusion of the interval, and a response translator which maps the contents of the counter onto the response surface."

In a GZE model (Gregson, 1993a) we have available a pacemaker in the terminal attractor which intermittently admits Γ trajectories, and this imposes a periodicity (though not a strictly fixed one) on inputs which are from a constant stimulus level and extended in cascade. It is possible to extend the analysis to cope with signals that themselves fluctuate periodically or quasi-periodically in their input levels.

Table 4.1: Notation

The symbols used are consistent with the treatment of nonlinear psychophysics in previous sources, and repeat some of chapter 1 for convenience.

Input signal level:	a_s
Starting point of first signal (or both, if in parallel):	t_0
Internal system variable initial level:	$Y_{0,1}$
Internal recursion before new input within Γ:	η
Output to next cascade:	Y_η
Duration of first signal in cascade steps:	t_1
Output to interstimulus interval:	$Y_{\eta,t1}$
Corresponding storage of first duration:	θ_1
Duration of interstimulus interval:	t_{isi}
Restabilisation within the interstimulus interval back to Y level at:	$Y_{0,2}$
Duration of second signal in cascade steps:	t_2
Corresponding storage of second duration:	θ_2
Eventual observed θ_1 after delay t_{d1}	$Y_{obs,1}$
Eventual observed θ_2 after delay t_{d2}	$Y_{obs,2}$

Y variables are complex throughout; a inputs are always real. The single channel model, Γ, as originally defined, involves the mappings

$$(stimulus)\ u \mapsto a, \quad \Gamma_{a,e,\eta}(Y(Re, Im)) \mapsto Y(Re) = Y_{obs}\ (response)$$

where Γ is the recursion, within one stage (cascade or superposition), for η steps, of the internal state variable Y,

$$Y_{(j+1)} = -a \cdot (Y_j - 1)(Y_j - ie)(Y_j + ie) \qquad i = \sqrt{-1}$$

for given starting values of Y, and fixed values of a, e. The stimulus input maps onto a, and the output is $Y(\mathrm{Re})$ at η.

The time points for two cascaded Γ recursions in series, where the stimulus to denote the continuing presence of a signal input is fixed at one level, are listed in Table 4.1. The component processes involving these parameters are Γ recursions within a time pulse, cascades of Γ recursions (Gregson, 1988, chap. 7), superposed Γ recursions for temporally extended inputs (Gregson, 1992a, chap. 10), restabilisation trajectories (chapter 3 here), and superposition by a GZE (Gregson, 1993a) to define the internal clock. That is, there are no new psychophysical mechanisms defined here for the current modelling task; all the theory already exists in the component parts, and the task is to see if combining these various mechanisms, whose mathematical properties are partly known in some detail, will exhibit the same diversity as that of the empirical results already known. Any theory that is robust and linear in its outputs with respect to its parameters t_1, t_2, t_{isi} is known a *priori* to be wrong.

If the two extended time intervals are presented in series then by definition they should be representable within the intervals by Γ superpositions or cascades in series, whereas if they are presented in parallel the superpositions or cascades of 2Γ (**Case 1** or **Case 2**) should be used. Under parallel input conditions it is known that the value of the psychophysical exponent is flattened considerably; there are capacity limits on the monitoring of parallel inputs. It is apparent from the results of Curtis and Rule (1977) that the serial and parallel cases have different dynamics; this shows in the failure of the Eisler (1981) parallel-clocks model. Allan (1992) gives some plots of psychometric functions for time duration whose forms resemble the distorted cumulative ogives produced by $\Gamma\ a \mapsto Y$ transformations. At the same time the differences between physical clock time and experienced short durations are not great, in the sense that ratios of durations are almost preserved in going from stimulus pairs to response pairs. If so-called power function scaling is used exponents of about 0.9 are found, with some variation over the range of stimulus durations used. Local discontinuities in psychophysical functions are predictable from the $\Gamma V1$ model, as shown in Gregson (1988), and irregularities in mappings will arise sufficiently using Γ V 7, as will be shown later. Eisler (1993) summarises some experimental evidence on this, but the statistical analysis is disjoint and linear, not nonlinear as used here. From this it follows that the magnitudes of time order errors (TOEs), expressed as percentages of stimulus inputs, are second-order, when they are detectable at all. They can reverse in sign (Needham, 1935) but positive TOEs seem relatively difficult to create in prospective estimation.

The total length of a duration comparison process between two internal representations θ_x of t_x, with the variable time delay δt between the end of the second presented duration and the response ($\theta_1 < \theta_2$ or $\theta_1 > \theta_2$ leading to corresponding ineqalities in Y_{obs}) is

$$t_1 + t_{isi} + t_2 + \delta t = T_\theta \qquad [4.1]$$

so for briefer notation

$$t_{d1} = t_{isi} + t_2 + \delta t \qquad [4.1a]$$

and

$$t_{d2} = \delta t \qquad [4.1b]$$

If the task is not to respond to two presented intervals, but for the subject to create the second interval as some multiple of the first, the $\delta t = 0$ and there have to be other changes in the internal representation of the second interval, not least because then $\theta_2 = Y_{obs,2}$. It follows from this that NPD can make different psychophysical predictions for the two tasks, prospective judgment and prospective reproduction.

The composition of the process during this T_θ in superposed or cascaded Γ trajectories is

$$\text{at lower level}: a \mapsto {}^{t1}\Gamma|Y_{0,1}, \eta, e \Rightarrow Y_{\eta,t1} \ (= \theta_1) \Rightarrow a*, Y_{\eta,t1} \Rightarrow a$$

$$\mapsto {}^{t2}\Gamma|Y*_{0,2}, \eta, e \Rightarrow Y_{\eta,t2} \ (= \theta_2) \qquad [4.2a]$$

$$\text{at higher level}: \quad {}^{td1}\Gamma(\theta_1) \Rightarrow Y_{obs,1} \qquad [4.2b]$$

$$\text{at higher level}: \quad {}^{td2}\Gamma(\theta_2) \Rightarrow Y_{obs,2} \qquad [4.2c]$$

where [4.2b]and [4.2c] are in parallel and terminate at the same point in real time. The (*) notation is used to represent the new transient system levels after partial restabilisation during the t_{isi}. Some studies have used a background level of stimulation (typically auditory) which serves as an interpolated stimulus within the t_{isi}. If this is represented in the model [4.2], then another a value, located at some t after t_1 and with a duration $t_3 \leq t_{isi}$ is also needed. The results of Needham (1935) suggest that variable t_{isi} with fixed null interpolation, and fixed t_{isi} with variable interpolated a are equivalent in their effects on the time-order error.

It may be noted that the idea of an internal trace which slowly decays, as the representation of the duration presented, is in the classical work of Lipps (1883), of Klien (1919) and of Köhler (1923), but the mathematical representation of such traces in scalars is quite different from Γ trajectories which are in complex variables Y by definition.

Sections through Response Surfaces

Given a system specification, in terms of structure and a set of parameter values upon which it is currently conditioned, a theoretical response surface is generated. Then an experimental design is, if well-defined, a section hyperplane through that response surface. In interpreting experimental results already available, two problems stand out; to decide where the section hyperplane (for short, the 'cut') lies, and to decide the tolerance on the results in terms of matching to the local qualitative features of the response surface. Some cuts are intrinsically more informative than others; for example, using only the stimulus conditions defined by $t_1 = t_2$ appears to be of little use in generating interesting response surface data. A fully-defined response surface, in this context, is a function sufficiently of the system's internal variables, but can be multivalued at any point. It is this multivalued property existing in some regions, as a consequence of the nonlinear dynamics, which generates more uncertainty in observed outputs (that is, corresponds to response distributions with variance). Matching of data to theory requires therefore more than comparing the first moments of two distributions, as is done in linear regression modelling.

The objective here is to produce a model of within-subject behaviour; inter-subject differences due to age, culture, gender, personality, or learning and other demand characteristics of the estimation task, can be found (Eisler, 1993), but are not of first consideration here. It it were required to extend the model to cover between-subject differences, then a *priori* distributions would have to be written on the parameters e, η, ζ, where ζ is a scalar multiplier on the superposition cycle number k.

Data Properties suggesting Nonlinearity

Psychophysical investigations have attempted to find some regularities in the mappings from

$$f(t_1, t_2) \textit{ onto } g(\theta_1, \theta_2) \qquad [4.3]$$

and have discovered persistent inconsistencies from one study to another.

If $t_1 = t_2$ and $\theta_1 < \theta_2$ then a *negative time-order (TOE) error* by definition exists; this is the commonest finding. However, for very small t_x the TOE often becomes positive. Sometimes no TOE is observed, and Jamieson and Petrusic (1975) observed that the TOE vanished when feedback was repeatedly provided after each judgment. Such feedback is, of course, an additional interpolated input, and would need additional representation in an NPD model. Most recent results (Wearden and Ferrara, 1993) indicate that the TOE changes in sign as a function of the response 'short', 'equal', 'long' made to trials where the $t_1 - t_2$ is negative, nil, or positive; a nine-cell table is needed adequately to summarise the relationships in the data. Accordingly, any simulations incorporating response selection from three categories and using Γ theory would have to reproduce this table. Additionally, reversal of the slope of the plot of $\theta_1 - \theta_2$ with increasing background noise during t_{isi} has been noted.

The actual responses collected have been $\theta_1 - \theta_2$, or θ_1/θ_2, or confidence ratings associated with *same*, *different*, or with *greater*, *less*. As any responses degenerated to binary from possible continua of differences in θ_1, θ_2 can be trivially predicted, but the reverse is not possible, simulation will be restricted to $g()$ in [4.3].

Mappings to be Identified

The simplest and most direct attack on the simulation problem is to plot

$$(\theta_1 - \theta_2) = f(t_1, t_2) \quad \text{or } \theta_1/\theta_2 = f(t_1/t_2) \qquad [4.4]$$

for fixed $Y_0, e, \eta, \zeta, t_{isi}$. It is to be expected that e will be critical, but η and ζ less so. This implies that a series of response surfaces for [4.4] have to be examined, and specific cuts (usually planar) through them compared with reported experimental findings. The reversal of TOE sign in the region of very small $t_2 - t_1$ implies steep gradients in that part of the response surface. The disappearance of the TOE, and the corresponding mapping near to identity in [4.3] (up to similarity transformations for scale units employed) for very long t_{isi} requires a separate series of response surfaces with t_{isi} as the parameter of the series.

First Model

The rather unusual course of briefly considering models which do not work, as well as models which do, is followed in this treatment, for illustration of what precautions should be taken. The first two models here are strictly wrong, and can be shown to be wrong in some critical detail when comparison with real data properties is made. The last two are more hopeful and lead to the model summarised eventually in Figure 3.7 . The stored representations of the outputs at t_1 and at $t_1 + t_{isi} + t_2$ are respectively θ_1 and θ_2. These are defined to be themselves complex variables of the form

$$\theta = complex(\sum_{j=1}^{t} Y_{\eta,j}(\text{Re}), \Xi) \quad [4.5]$$

the underlying idea here is that the storage of the θ at one level higher than the original trajectories is itself on Γ attractor dynamics, preserving the dual information implicit in the output of a cascade of a duration clocked on a terminal attractor whose periodicity effectively fixes η for us. In jumping levels, the $Y(\text{Im})$ to the outputs t_1, and t_2 is lost, and t is mapped onto $\zeta\Xi$, where ζ is a scaling constant, as the Im parts (Ξ) of a higher-level Γ recursion, because it is treated as the modifier and not as the t signal carrier at that level.

The parameters to be specified to set up the model are $a, t_1, t_{isi}, t_2, c, \eta, \zeta$, and the observable consequences of the model's evolution in time, in this parameter space, constitute the generation of a θ_1, θ_2 response surface. If there is no distortion of clock time in psychophysics (that is, the exponent of a fitted power function is nearly everywhere unity) then a map of the ratio of time durations, for $min < t_1 < max$, and $min < t_2 < max$ will be a radial family of straight lines inclined for $t_1 = t_2$ at 45°. Using a model with $\zeta = .00001$ we get Figures 4.1 and 4.2 , which are quite different for small and large time durations.

This first model is interesting but wrong; it fails to reflect two properties of real data: (a) That the exponent of a fitted $log(Y_{1,obs}/Y_{2,obs}) = b \cdot log(t_1/t_2)$ should lie around 0.9, when in fact it is almost 1.0 for $t_1 = t_2$. (b) The distortion from the nearly linear map of t_1/t_2 to the observed convoluted form of the output map accurs *after* the intermediate (θ_1/θ_2) created at the offsets of t_1 and t_2 which are before higher-level storage. But the results of Brown, Stubbs and West (1992) show that the transformations from (θ_1/θ_2) to observable delayed output $(f(Y_{obs}))$ should be minimal; the nonlinear form of the response surface should already be generated at the θ phase. This indicates that the cascading within a t_x duration is wrongly parameterised, and that the higher-level stage transformation [4.5] can temporarily be dropped from modelling. This brings us to a second model.

Second Model

In the first model very short η values were used, and relative high Y_0 and a_t. However, the cascade equation within time periods, from step k to $k+1$, $(0 \leq k \leq t_x)$ was written as

$$a_{k+1} = c_1 Y_k(\text{Re}) + c_2 \quad [4.6]$$

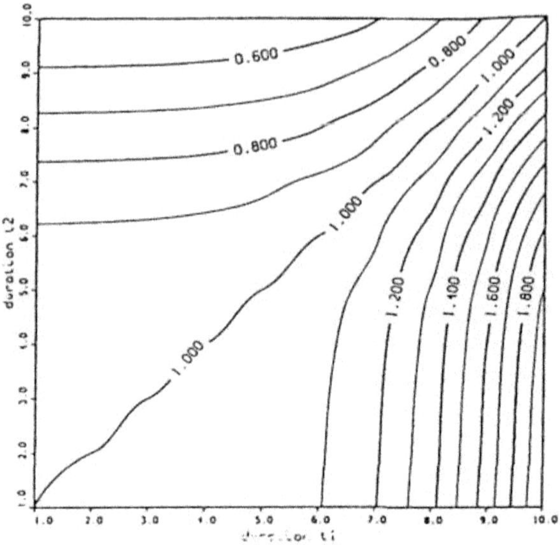

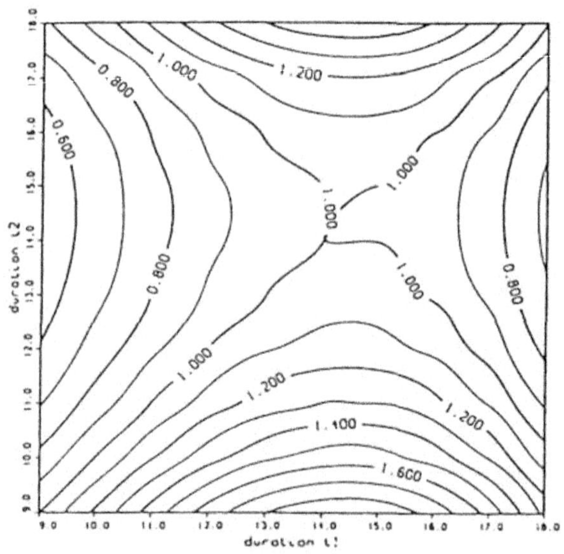

Figures 4.1 and 4.2:
Time Duration Ratios at Output for a First Model;
Low and Higher Stimulus Ranges

The constants c_1, c_2 in [4.6] are to keep a_{k+1} within the range 2.01 – 5.5, given that by definition $0 \leq Y(\text{Re}) \leq 1$; [4.6] is independent of the duration increments as such.

In the second model the possibility, that the internal variable's decay within steps η long is diminished as a proportionate function of the duration's evolution, is considered. That is, the strength of the internal variable Y is marginally linearly incremented by the pacing mechanism:

$$a_{k+1} = c_1 Y_k(\text{Re}) + \zeta k + c_2 \qquad [4.7]$$

This idea is reminiscent of a trace theory advanced by Wickelgren (1972), which also requires dual encoding for both strength and resistance to extinction.

The negative TOE can be reproduced with these assumptions, and the ratio θ_1/θ_2 is less than t_1/t_2 when $t_1 \simeq t_2$. However there is an indeterminacy in this situation, which also casts doubt on data collected by matching t_1, as presented, to a new $\psi t_1 = t_2$ as the prospective reproduction of t_1, which is the paradigm used by Eisler(1993). As

$$log(\theta_1/\theta_2) = b \cdot log(t_1/t_2) \qquad [4.8]$$

if the purpose of the experiment is to estimate the exponent b of an assumed power law in duration psychophysics, then [4.8] is ambiguous when $t_1 = t_2$, because the r.h.s. becomes zero.

The value of a_{isi} has also been set very high, at 5.41, to achieve rapid restabilisation of Y_0; this parameter was chosen for the Γ restabilisation trajectories. It appears in this model that the length of the i.s.i. could be crucial for determining the TOE, which does agree in part with results summarised by Hellström (1985).

Third Model

The assumption ζk in [4.7] looks like cheating; it is building into the process exactly what one wishes to get out, even though the consequences of adding this incremental term does depends for its effect on a_0, e and η. Ideally the model should use only Γ trajectory properties without ancillary mechanisms to produce the observable psychophysics. To try this we drop the cascade assumptions of [4.6] and [4.7] and use instead serial superposition of a fixed input, as modelled in Gregson (1992a, chap. 10).

Now

$$\forall k, \; a_k = a_t \qquad [4.9]$$

Table 4.2
Parameter Settings for an example with the Third Model

$$1 \le t_1 \le 10, \qquad 1 \le t_2 \le 10$$

$$t_{isi} = 2, \qquad e_t = .14, \qquad e_{isi} = .12 \qquad \zeta = .01$$

$$a_t = 4.0, \qquad a_{isi} = 5.4, \qquad \eta_t = 15, \qquad \eta_{isi} = 5$$

Again there is a recursion on steps $1 \le k \le t_x$,

$$Y_k(\mathrm{Im}) = \zeta k \qquad [4.10]$$

so that

$$\Gamma_k = f(a, \eta_t, Y_k) \qquad [4.11]$$

and in storing representations of a sustained input at the θ level the $Y_k(\mathrm{Re})$ are accumulated by

$$\theta_t = complex(\sum_{k=1}^{t} Y_k(\mathrm{Re}), \zeta t) \qquad [4.12]$$

This model is so far the most interesting, so some detailed parameter settings are given in Table 4.2. These were obtained partly from knowledge of the Γ system's dynamics explored in previous psychophysical studies not connected with time duration perception nor using temporal stimuli, and partly by exploring the parameter space on a sampling basis in the neighbourhood region within which the given values lie. Taking this approach explicitly assumes, by Occam's razor, that time estimation is not a unique process but one mediated by the same dynamics as all other sensory inputs presently explored, and that parameter settings which induce stable but nearly chaotic trajectories should be expected here, particularly in the restabilisation interstimulus interval.

An idea explored here is that the unperturbed quasi-stationary frequency of the carrier dynamics provided by terminal attractors is higher than that during a sustained sensory input, the stimulus "pulls down" the dynamics into a slower-paced process below the bifurcation regions; at the same time the cumulative input shifts the $Y(\mathrm{Im})$ output values to θ_x. This attenuates the rate of attraction of the complex Y trajectory onto the real line. To put this another way, the rate of convergence onto stability is reduced. There is a necessary interplay between a_t, e_t, η_t, ζ and t_x in all these dynamics which eventually fixes θ_x and due to the very high a_{isi} in restabilisation a marked order asymmetry between $g(\theta_1, \theta_2)$ and $g(\theta_2, \theta_1)$ arises[3]. It is found that $\partial\theta/\partial\zeta$ is negligible over a wide range of ζ in[4.10].

Interestingly, this model only produces plausible maps of time duration ratio differences in a very narrow parameter range for $a_, a_{isi}$, and the negative TOE again emerges. To summarise observed and predicted properties we need both the response surfaces shown here, θ_1/θ_2 for comparison with psychophysical scaling and TOEs, and $log(\theta_1/\theta_2) - log(t_1/t_2)$ for comparison with Vierordt's law. The ratios and differences of outputs are themselves complex variables, but the Im parts are second-order and can be dropped in generating response surfaces sufficiently from only the Re parts of θ_x. Figure 4.3 of ratio outputs (θ_1/θ_2) shows the displacement of the contour for ratios of equality (=1.0) downwards; this corresponds to a negative TOE, because now $t_1 > t_2$ is necessary to induce $\theta_1 = \theta_2$. It is noted that the θ_1/θ_2 contours around very short t_1, t_2 have irregularities in them; this particular departure from linearity in the contours is better seen in Figure 4.4 where the differences $log(\theta_1/\theta_2) - log(t_1/t_2)$ are plotted. (The base of the logs is immaterial.)

The plot of ratio outputs does cast some light on three reported properties simultaneously; (a) the irregularities[4] of the contours near to small t_1, t_2 coordinates is necessary to produce the nonlinearities which Eisler (1993) models with straight line segments and breaks in an assumed power law representation, (b) the displacement of the contour for $\theta_1/\theta_2 = 1$ downwards does predict the commonest negative TOEs, and (c) for matching two equal t_x nearly veridical θ ratios should be produced. That is, all three phenomena are interconnected consequences of one nonlinear process, though there may be additional pro-

---

[3] The treatment here of the relative Γ dynamics of sustained inputs and of restabilisation between such inputs is not dissimilar from some results of Freeman and of Haken which are explictly acknowledged in Gregson (1988); namely, that stimuli in olfaction pull chaotic dynamics down into attractors whose trajectories in their phase space encode qualitative stimulus information, and that the brain can only exhibit learning and relearning capacities by operating transiently into and out of chaos.

[4] It should be noted that the graphed irregularities are the result of DISSPLA fitting continuous contours to a lattice of points; if Eisler (1993) had used cubic splines instead of broken line segments to fit her psychophysical plots a different interpretation would have eventuated.

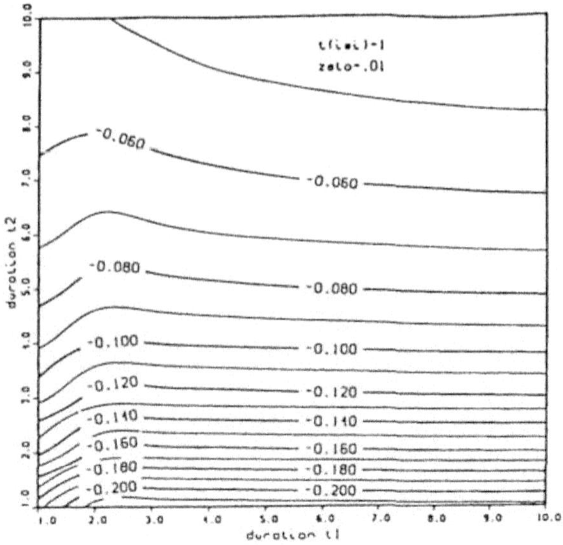

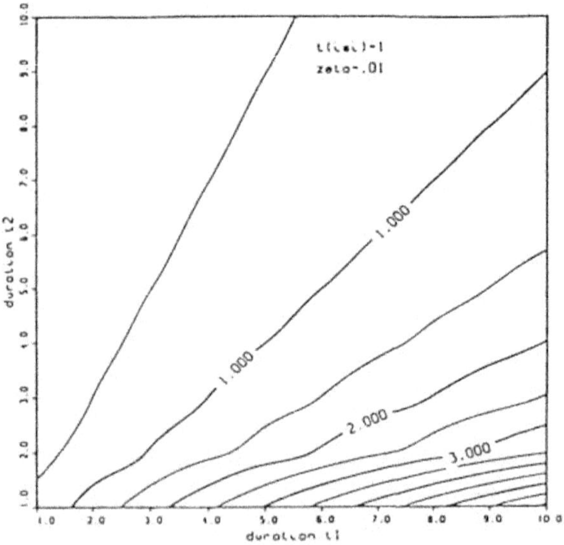

Figures 4.3 and 4.4:
Log Ratio Differences and Output Ratios for the Third Model

cesses also sumperimposed at the response selection stage from $\theta \mapsto Y_{obs}(\text{Re})$. If this final $Y_{obs}(\text{Re})$ output is taken as a numerical rating then the extensive data tabled by Nakajima, Nishimura and Teranishi (1988) are relevant for comparison. The parameter ζ has only a second-order effect; increasing it reduces the TOE for $t_1 = t_2$ to zero, but leaves the contours almost unchanged in their relative shapes. The plot of log ratio differences may have some bearing on the generation of Vierordt's law; if the reversal of slope of contours for $t_1 \simeq 2$ implies that very short t_x are *relatively* overestimated, then a fourth phenomenon is also tied into the NPD repesentation.

It also follows that taking a cut through the response surface up the leading diagonal, which is what is done if only matching of $t_{\simeq}t_2$ is the task used, is relatively almost minimally uninformative about the response surface's shape. In this model there is no provision for higher-level trace decay of the θ_x which are taken as very stable within the time span T_θ. In, but only in, this feature there are correspondences with Eisler's (1981) parallel clocks idea, because θ_1 generated at t_1 has to be preserved until θ_2 is also generated and becomes available for comparison judgments. The mechanisms of comparison θ_1/θ_2 and $\theta_1 - \theta_2$ (as stored after delays t_{dx}) are taken as almost instantaneous when made and therefore approximately linear.

So far, in this argument, no mention of $var(\theta_x)$ generation has been introduced; for interim consideration of any matching between Γ theory and reported average responses upon which most experimental generalisations are apparently based there are no data on variances, apart from those of Nakajima *et al* (1988), except from intersubject comparisons. Such variances can be very small, expressed as σ/μ.

Fourth Model

Whilst the third model shows features which are qualitatively what are being sought, there remains the question, what changes will reintroducing the higher-level trajectories of θ_x terms produce ? Such changes ought to be second-order, but at the same time if they exist should accentuate some features of the response surfaces, when $f(Y_{obs})$ are plotted instead of $f(\theta)$. The assumption of [4.8] is retained, and the arguments underlying [4.1] and [4.2] reinstated. The parameter values of Table 4.2 and [4.12] are retained, for the generation of θ_1, θ_2, and for the higher level processes [4.1] and [4.2] are used.

At the higher level, there are no superposed inputs on θ_x, so Γ cascading is the only mechanism available from the pre-defined alternatives already listed in the introduction. In this fourth model, therefore, we employ Γ superposition at the lower level, and Γ cascades at the higher level. The parameters at the higher level chosen for this illustrative example are $\eta = 5, e = .35, \delta t = 5$, and

$$a_k = \theta_{k-1}(Re) + 2.01 \qquad [4.13]$$

This is for a $Y(Re)$ creation over t_{dx} cascade steps, implying perseveration of a Γ trajectory with recursive regeneration by $Y_{k-1} \mapsto a_k$. In this sense the third model could be have been defined, for coherence, as $Y_{obs,x} = \theta_x$, bypassing [4.13]. The results of plotting $Y_{obs,1}/Y_{obs,2}$ and $log(Y_{obs,1}/Y_{obs,2}) - log(t_1/t_2)$ are shown in Figures 4.5 and 4.6. If the t_x were very long then [4.13] would not work without modification.

It is seen that, given the parameter values used, the consequences of putting the terms θ_1 and θ_2 into cascade at the higher (storage) level are second-order for their effect on ratio outputs. There are however different and more marked consequences if the $log(Y_{obs,1}/(Y_{obs,2}) - log(t_1/t_2)$ response surface is plotted.

The storage cascading flattens the t_1/t_2 surface, so that as Nakajima *et al* (1988) observed, t_1/t_2, $(t_1 \neq t_2)$ ratios converge towards 1:1, and some nonlinearity arises in the contours plotted for $t_1 << t_2$. The negative TOE is preserved, even accentuated. In this sense, storage delay at a higher level of internal dynamics does not much alter the observable psychophysical mappings at ratio level, which is consistent with Brown *et al* (1992). However, the effects on the $log(Y_{obs,1}/(Y_{obs,2}) - log(t_1/t_2)$ response surface are quite different; now the contours run from negative to positive, so the surface implies that only in the region $t_1 \simeq t_2$ would we expect to get nearly veridical judgments ($.9 \leq b \leq 1$ in power law scaling). Contour irregularities emerge again for low t_1, and the supposed breaks found from fitting straight line segments[5] to data may be generated or aggravated at this level.

Filled versus Empty Time Duration Estimates; a Pseudoproblem ?

The *filled duration illusion* is the name given to the commonly reported phenomenon, that durations filled with stimulus inputs or distractions are often judged longer than physically equal empty durations. Zakay, Nitzam and

[5] The existence of a multiplicity of apparent discontinuities in the ψ/ϕ plots does not necessarily indicate that series of temporal oscillators are mediating perception. In fact one attractor whose dynamics lie within its bifurcation series, coupled to another such attractor, can exhibit such multiplicities. The error of supposing that it is necessary to postulate "an extensive system of parallel pacemakers" is found in Treisman, Cook, Naish and MacCrone (1994).

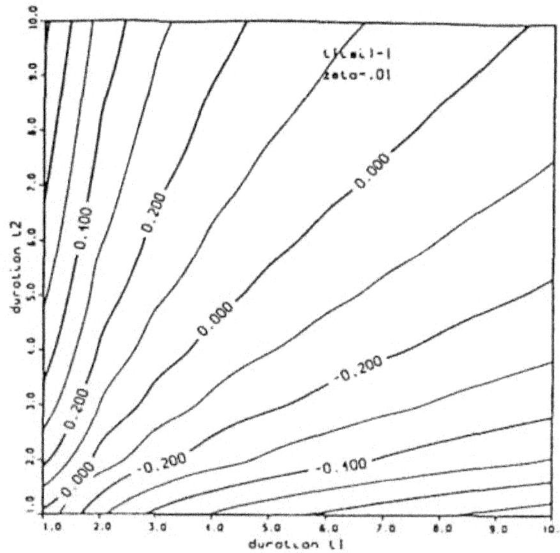

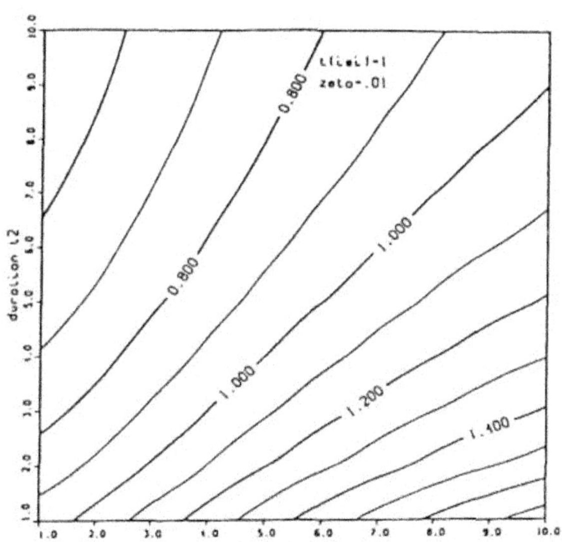

Figures 4.5 and 4.6:

Log Ratio Differences and Output Ratios for the Fourth Model

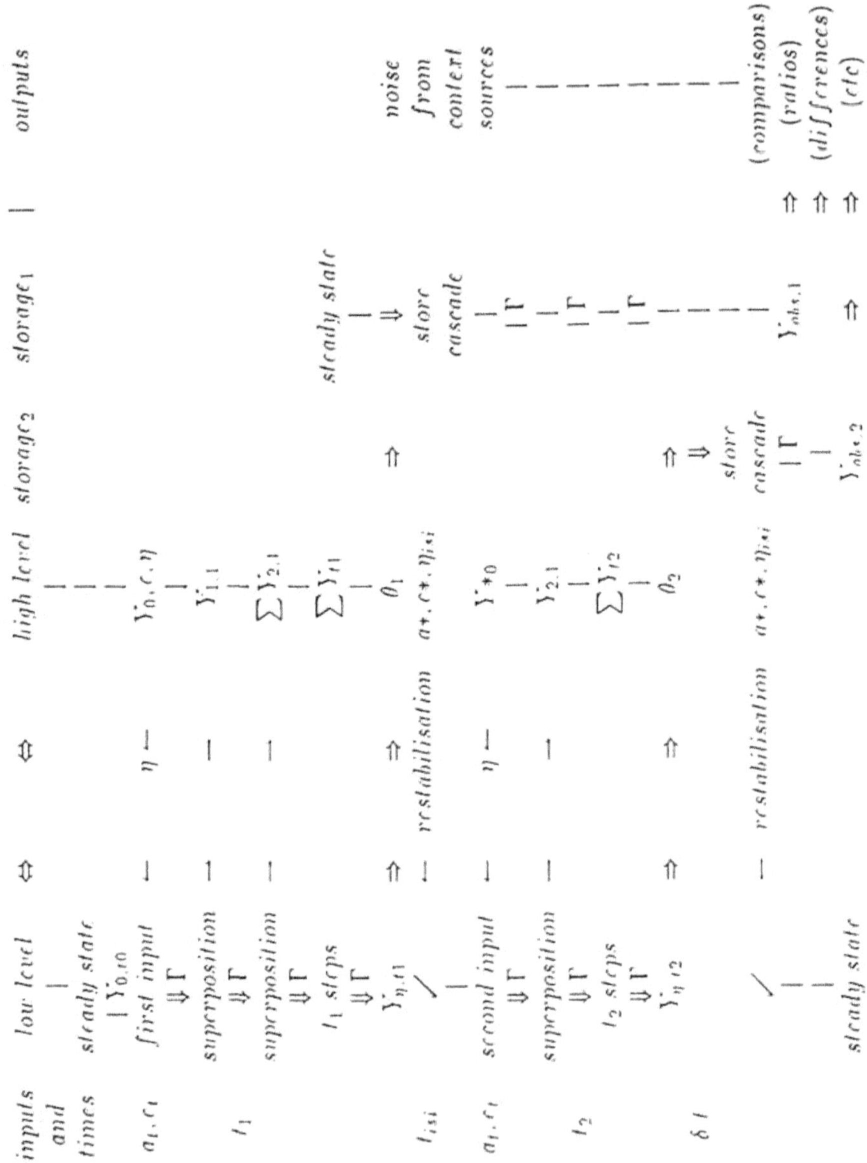

Figure 4.7: Prospective Serial Estimation; Dynamics of System Construction from NPD Components for the Fourth Model

Glicksohn (1983) is a typical study on this problem. However, there are some ambiguities of definition in the literature, and conflicts in results which may be the consequences of methodological variations. Zakay *et al* (1983) found that duration estimations, made by reproduction in a prospective paradigm, were a decreasing function of task difficulty (interpolated verbal tasks were given) and empty intervals were judged longer than filled intervals. As Zakay (1990) comments "the whole notion of empty time is problematic". In very long intervals there is there is little filled duration effect reported, and the more difficult the discrimination of temporal duration between two nearly equal intervals, one filled and the other empty of material introduced by the experimenter, the bigger the illusion. It is suggested by some writers that the illusion is the consequence of nontemporal factors taking over to meet the demand characteristics of the imposed but almost intractable temporal discrimination task.

In this purely theoretical note we examine the results from one nonlinear Γ superposed and cascade interval model for prospective psychophysical duration ratio judgments. It is expedient first to reiterate that the ideas of 'empty' and 'filled' are imprecise, and are rooted in the stimulus conditions, including instructions, given to subjects, and not in a differentiation – internal to the subject – of the content of the representations of the intervals during the durations. This point is made by previous writers when they note that though the experimenter may not fill an interval, we have no nonintrusive way of telling what the subject is putting into it, by way of musings, fantasies, calculations or recollections. If there is a sharp diffierence in filled and empty stimulus conditions or tasks, it cannot be confirmed that that sharp distinction is internally represented, though continuously varying differences of degree may be noticed or internally encoded.

In this present deliberately restricted case we examine only one variable, mapping onto a_t, which is the level of sustained uniform stimulation within a predetermined duration. Call this stimulus level U_t, as distinct from U_0 which is the stimulus display level immediately before a duration is begun. For example, an observer may be asked to estimate the duration within which a pattern appears on a VDU screen, the screen being otherwise blank. The brightness or complexity of the pattern is then U_t, and is potentially variable but in the simplest experiment is fixed during all the durations to be compared.

Now $U_t \mapsto^m a_t$ in the Γ superposition series representing the input of duration t_x. Suppose that there is a series of replications r in which $U_{t,r}$ is fixed within any replication but varies across replications. As $\mapsto^m$ is affine, a_t also varies. The level of U_t is the degree of filling, but a_t is never nothing (by Γ

dynamics), and a_t can be saturated at an upper level (say $a_t = 5$). To see what the consequences of varying U_t while keeping all the other system parameters fixed, the fourth model from the previous analyses is used. The graphs of $\theta_1/\theta_2 - t_1/t_2$ are plotted as contour maps for fixed a_t. Figure 4.8, which gives contours for t_1/t_2 only, is given as a starting point; the remaining graphs are then maps of the degree of distortions of the contours in Figure 4.8, both in loss of linearity and in their associated values. If there were no transformations from $t_x \rightarrow \theta_x$ in the sense of losing veridical reproduction of ratios then all the figures 4.9 through 4.14 would be blank, because the surfaces would be flat and everywhere zero.

If the psychophysical transformations were linear but preserved equality, so that

$$Iff\ t_1 = t_2\ then\ \theta_1 = \theta_2\ then\ Y_{obs,1} = Y_{obs,2}$$

then the contours in the error maps of Figures 4.9 through 4.14 would be straight lines with a radial pattern and zero on the leading diagonal $t_1 = t_2$.

It is known that judgments of t_1/t_2 or of $t_1/(t_1 + t_2)$ can be made quite accurately for t_x ratios that are not large (< 10), and show some irregularites and deviate below for very small t_x. This error pattern is reviewed in the previous treatment. All these qualitative features, and the instability of results from one published study to another, should emerge as properties of the theoretical dynamics, as a consequence of varying a_t, though secondary variation in other parameters might also be involved.

The series of graphs $a_t = 2.6, 3.0, 4.0, 4.05, 4.3, 4.6$ in Figures 4.9 through 4.14 illustrates the qualitative changes in the difference response surfaces. At the lowest a_t cases, the error contours cross the diagonal $t_1 = t_2$ from low t_x, and are for small error values. Note that the errors are always negative, this means that $\theta_1/\theta_2 < t_1/t_2$, always, or in other words values are flattened but small relative durations with low inputs are preserved.

If a_t is gradually increased the differences become bigger, and the difference surface rolls off more sharply, and to a greater degree, for low t_2. The changes across the range $3.0 < a_t < 4.0$, which is a range typical of Γ simulations in a diversity of psychophysical processes, are small. That is, the dynamics are fairly stable, with the irregularities for low t_1, t_2 being more obvious. This confirms that $\theta_x = t_x^b$ is not valid, and has to be roughly patched up by allowing b to jump at some values of t_x, as in Eisler's breaks. It is seen that the error values over the whole surface increase on average with increasing a_t. That should imply that filled intervals (high U_t is one sort of filling) may be more

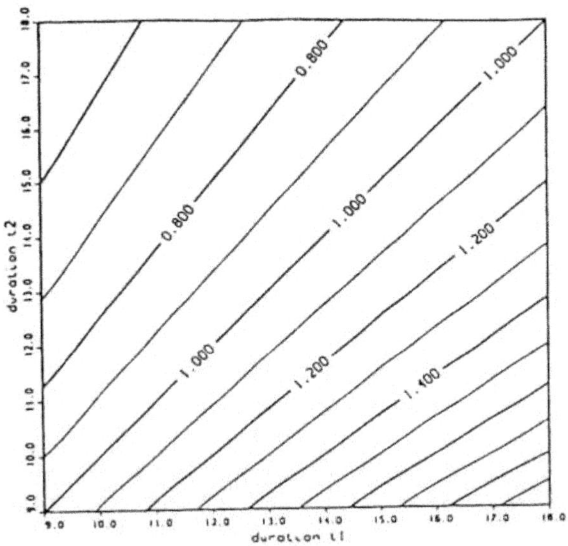

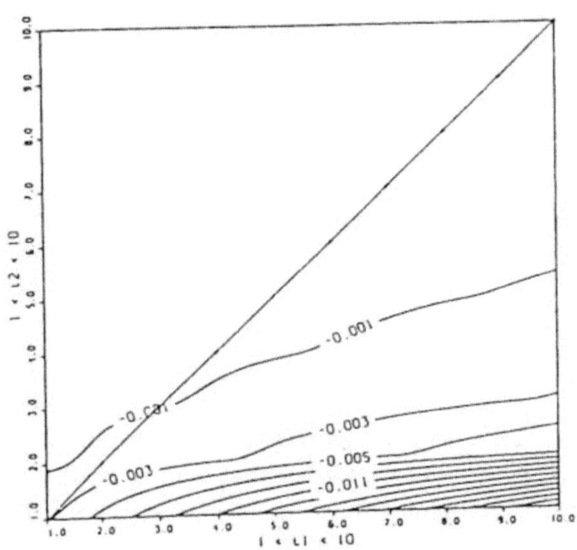

Figures 4.8 and 4.9

underestimated, which is what Zakay *et al* (1983) report.

About $a_t = 4$ things begin to happen, which is why both $a_t = 4.0$ and $a_t = 4.05$ are shown. It is of interest that by trial and error $a_t = 4.0$ was chosen to be used in the previous simulations. The rapid changes in topography (the topology is invariant) are easily seen if we focus only on the contour intersections with the $t_1 = t_2$ diagonal marked on all the graphs. As a_t is increased further, in the next example at $a_t = 4.3$, the emergence of contours independent of t_1 for large t_2 begin to dominate the picture; at first for higher t_1. That is, the degree of bias on θ_1/θ_2 is independent of t_1, but the errors are still very small unless t_2 is very small. This is reminiscent of Vierordt's law.

In the last graph, $a_t = 4.6$, which is about the limit before chaotic explosion, given the settings of the other parameters (ϵ, η, ζ) in the Γ system, the dominance of horizontal contours is almost total, with exceptions for $t_2 < 2$, and the rolloff for $t_2 < 2$ is steep, but in fact to lower numerical values than for $a_t = 4.3$. Considered over the whole t_1, t_2 response surface map, there are qualitative changes as a_t is moved over the dynamically stable range available, and the changes are not everywhere monotone on a_t. We expect that irregularities will emerge in the neighbourhood of the first Γ trajectories' bifurcations, but it is still interesting to trace out the implications of nonlinear dynamics in maps of disparities between ratios which themselves are created in a complicated pair of sequences in real time.

U_t		a_t		θ_t		Y_{obs}
empty		*continuously*		*qualitative*		*new*
versus	$\Rightarrow$	*varying*	$\Rightarrow$	*shifts*	$\Rightarrow$	*contrasts*
filled		a_t		*in*		*in*
by	$\Rightarrow$	*over two*	$\Rightarrow$	*error*	$\Rightarrow$	*outputs*
contrasted		*separate*		*surface*		*called*
inputs		*ranges*		*topography*		*f.d.illusion*

The total situation as an input-output mapping is represented in the flow diagram $U_t \rightarrow a_t \rightarrow \theta$ which suggests that the contrast between empty and filled at input is converted into a shift in the dynamics at output, which may then sometimes give rise to the differences interpreted as the filled duration illusion. There is no necessity for the point at which 'empty' becomes 'filled' to match exactly the range of the dynamics with rapid qualitative shifts, as are shown to occur in $4.0 < a_t < 4.05$ for the example graphed. Consequently, it does not

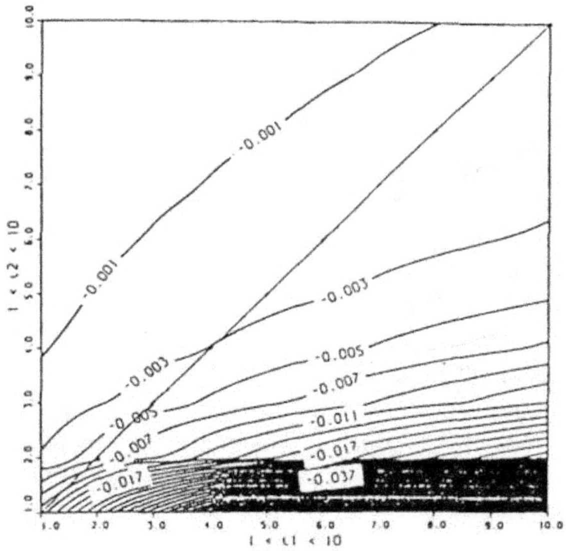

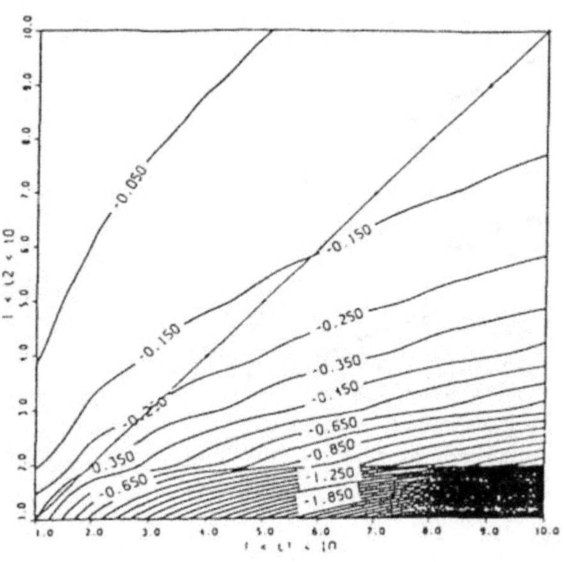

Figures 4.10 and 4.11

always follow that the illusion will be generated, nor that it will be invariant in its direction.

Discussion

If Γ theory in the first model, with $\theta = \text{complex}(\sum_{j=1}^{t} Y_{\eta,j}(\text{Re}), \Xi)$, is used as a representation of time duration psychophysics, it is only in a very narrow window of the system's parameter space that response surface cuts which correspond to observable data can be reproduced. The parameter window is wider in the third and fourth models. This is not a disadvantage, on the contrary if the parameter settings of η representing the pacing of the inner Zeitgeber dynamics, and of t_1, t_2 representing real time durations, are critical, then the generation of such phenomena as Vierordt's Law, and, for equal t_1 and t_2, the nearly linear psychophysical mapping (exponent values b; $.9 < b < 1.0$) emerge but only under restricted conditions. The lack of correspondence between experimental studies in determinations of the TOE and the indifference interval implies that the phenomena are not robust over a wide range of contextual parameter settings.

Likewise, the role of the duration of the interstimulus interval (t_{isi}), or of background noise signals in that interval, has to be represented in the settings of a_s and ζ in controlling the decay of θ terms after the offset points of t_1 and t_2. It appears that η is not very critical in forming the topology of the response surfaces, but e_t and e_s only admit non-chaotic dynamics if held low ($< .15$).

In the only viable (third and fourth) alternative models considered, the results of prospective time duration experiments which generally include second-order departures from veridical quantitative perception appear to emerge as natural properties of the NPD. In fact the selection of parameter values to produce the response surface maps is not extremely critical; the values of a, e, η chosen in each component, as in [4.2] and pictured in Figure 4.7, are substantially the same as used in various previous and diverse psychophysical problems (for example, Gregson, 1993c). It may be re-emphasised that if time durations in parallel, and not in series, were involved, then Figure 4.7 would need additional columns on the left hand side for each input, and cross links between those columns at each superposition step for the λ parameters in $n\Gamma$ (see Gregson, 1992a). The most important finding of this theoretical exploration, drawing on the summed properties of published response data, is that various time perception phenomena are intrinsically interlocked, so that in generating any one such as a negative TOE, other properties of the response surface emerge, and the nonveridicality of the psychophysics is indeed limited.

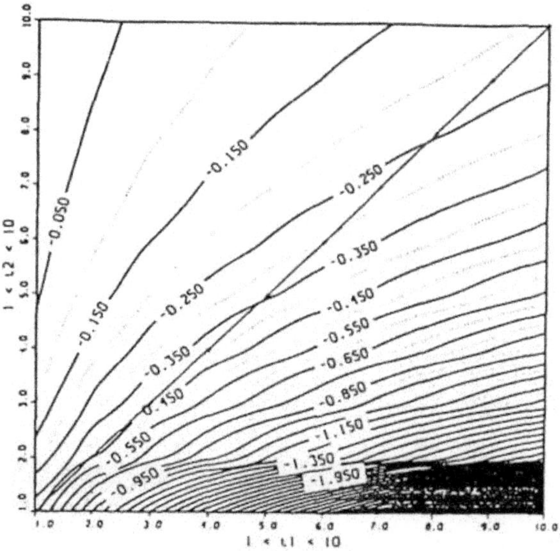

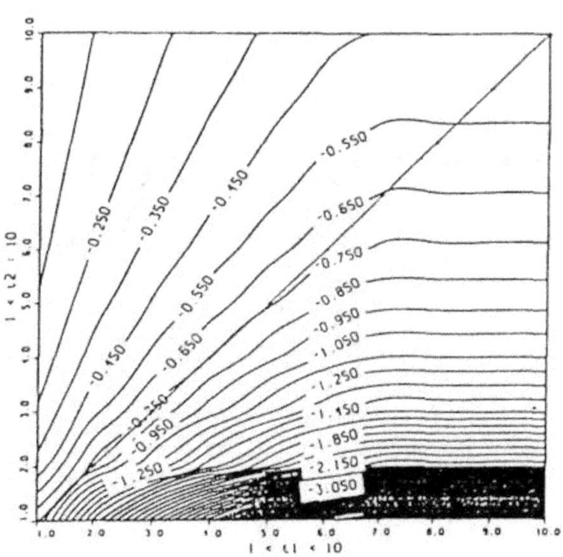

Figures 4.12 and 4.13

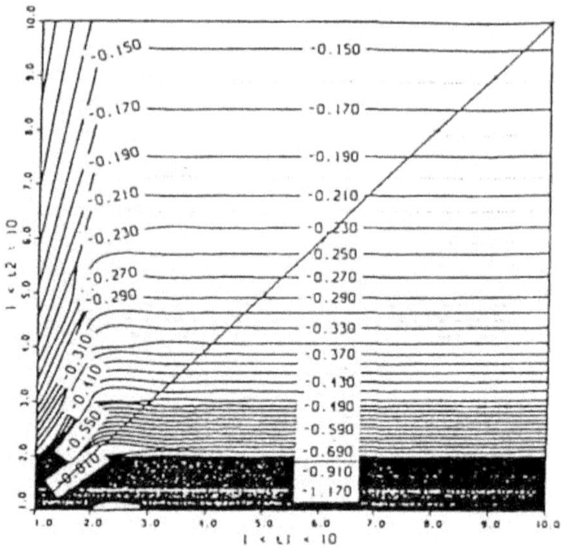

Figure 4.14

But a wider range of cuts than has often been used is demanded if the topic is to be fundamentally understood.

The number of freely-defined parameters in the Γ trajectory recursions, and couplings between trajectories, is not a measure of the degrees of freedom of the models if their fit to empirical response surfaces is to be considered. To define the effective dimensionality of the NPD at a point P in the parameter space, it is necessary given the set of m parameters $\{\pi\}_P = a_P, e_P, \eta_P, \zeta_P, \ldots$ to know the Jacobian

$$J_{m|P} = \frac{\partial Y}{\partial \pi_1}, \ldots, \frac{\partial Y}{\partial \pi_m} \qquad [4.14]$$

which as Y is complex, is $2 \times m$. The cardinality, C, of the subset of non-null terms of $J_{m|P}$ gives the local dimensionality at P. If the psychophysics is an Andronov stable system then C is low $<< 2m$, and the stability of the theory should match that of the actual psychophysical system being modelled.

A metatheoretical point of interest has been raised by some writers (Zakay, 1990), in noting a parallel between Heisenbergian indeteminacy in quantum physics, and indeterminacy in the experimental psychology of time. If the process of observing a phenomenon is itself intrinsically a part of the phenomenon, then the phenomenon cannot be observed without altering it. It is suggested that in asking observers to estimate time we induce changes in the way time durations themselves are monitored, whether or not we postulate internal Zeitgeber mechanisms, as we do here by drawing on GZE dynamics.

Where can and do we go next ? We have shown that a process model ties into one interdependent framework, via response surface topologies, the outcome phenomena that are treated as separate topics in outcome modelling via linear psychophysics. Yet the NPD ideas are no use for memory, or pattern recoognition, without augmentation; the best that they can do is to serve as input to networks. At the same time they are pictures of hypothetical attractor dynamics and so are starting point dependent. If they are to be incorporated in learning processes then we have to know how they would be used to kick networks out of false local basins of attraction. In the senses identified by Freeman and independently by Haken the Γ recursion can run at chaotic levels and can be pulled down by inputs into limit cycling, which is what we need for some structure encoding. There is now some evidence that chaos-based learning (by which is meant processes which have some access to chaotic dynamics, but not permanently in chaos) introduces some apparent randomness into networks, at a level which is necessary and sufficient, and in this respect it differs from annealing which always introduces the same fixed schedule randomness. It is known that annealing and backpropagation do not work as efficient regulators of cognitive processes, but coupling NPD with networks could create the situation where, given a target outcome of behaviour, searching amongst transformations which might generate the required output, and these transformations are a closed set of nonlinear trajectories.

5 Effects of Phase Space changes in the Dynamics of Lattices and Noise in Fields

Faraday in 1821 experimentally identified the magnetic field around a wire carrying a current of electricity, and Maxwell later set up the mathematics to describe formally the properties of such a field (Luchins and Luchins, 1965). In the sense it is used here, the term *field* owes its origin to Faraday's magnetic investigations. Mach (1866) employed a form of field theory, partly derived from Fechnerian psychophysics, to describe phenomena in the perception of contours and brightness gradients. Field theory held unchallenged sway in physics until the advent of quantum theory in the twentieth century, and left its mark in Gestalt psychology following its adoption in a discursive qualitative form by Köhler (1920).

To explore the metatheoretical problems with which we are here concerned, examples or definitions of *field, neighbourhood, lattice* and *network* are needed. Brief working definitions are set out in Table 5.1. A field is everywhere continuous in its spatial coordinates, and every point in it has a potential value, which can be expressed as a vector of forces, or a gradient of potential levels locally. A lattice may be taken, if it is sufficiently dense, as a discrete mapping of a field, and the process of diffusion from a point location into its neighbourhood can be represented as a probability distribution onto adjacent points in space and in time. In field theory this is a continuous process which in one or two dimensions may generate something like the generalisation gradients found in conditioning experiments. In a lattice it involves a spread onto adjacent pixels, and something analogous is involved in the theory of crystal structures, where inclusions may diffuse through the interstices of a lattice. A network is a lattice that is internally connected to some degree, not necessarily completely, but at least in such a way that by recursive diffusion a point input may spread in its consequences to a region of the lattice. The formal theorems on connectedness in lattices are in graph theory (Harary, 1972), and a treatment which puts recursive lattices into a wider mathematical context is given by Grenander (1993, p. 407 et seq.). The mathematics of physical or chemical diffusion processes has a very extensive literature, but the assumptions which motivate such treatments do not exactly carry over into psychological problems.

Table 5.1: Working Definitions

FIELD

Everwhere continuous in x, y coordinates, definable in terms of local potential z, and its derivatives, dz/dx, dz/dy, d^2z/dx^2, d^2z/dy^2,.. (at least). Can be represented as a contour map.

NEIGHBOURHOOD

Either spatially or functionally a set of C-connected points which can be defined as nearer than some ϵ to the system's current reference point.

C is some arbitrary function which creates a signed graph in, and only in, that neighbourhood.

LATTICE

A set P of points (or nodes or pixels) defined on an evenly-spaced bivariate discrete array, usually rectilinear or hexagonal.

NETWORK

A set of lattice points P together with interpoint connections C which exhibits a degree of connectivity, either locally in an ϵ-neighbourhood or globally. The connections enable information to be moved between points.

SPLINE FITTING

A method, given a lattice with z at each point, of deriving a continuous map, which in the limit, when the lattice is sufficiently dense, resembles a field contour plot.

An $(n \times n)\Gamma^k$ process is a cascaded series of latticed inputs a_{xy} which through cross-coupling become networks evolving in time. The evolution is orthogonal to the plane of the bivariate lattices. Its evolution can thus be drawn as a series of field plots, one at each cascade step.

Diffusion replaces a point source potential value by a distribution of potential values about the spatial coordinates of the original point source; it is not necessary that the potential should be monotone decreasing in all directions away from the point source, but the bounds of the diffusion region enclose a region, or set of pixels, which is finite in finite time after input to the source. The values of potential are not constant over time within this region, and will decay to zero or to background stability levels eventually. In $(n \times n)\Gamma$ networks this could happen quite quickly, depending critically on the cascade between-step mappings $Y_k(Re) \mapsto a_{(k+1)}$.

For illustration[1], consider the high school elementary laboratory demonstration where one bar magnet is placed on a sheet of glass, and iron filings are scattered on the glass. These filings are themselves bar-shaped, and each one takes up an orientation in a fixed location. The resultant total pattern of these filings illustrates what are sometimes called *lines of force*. These lines resemble closed contours looping through the two poles and axis of the magnet. To describe such a situation we could list exhaustively the coordinates (x, y) of the ends of every filing, δ long, as $((x, y), (x + \delta x, y + \delta y))$ after they have come to rest. The filings come to rest as the resultant of three forces, the two poles of the bar magnet and the friction between the filings and the glass surface upon which they rest. Each little filing is itself a bar magnet. Its orientation on the glass surface reflects the resultant of the forces acting on it. To replace the exhaustive catalogue of the filings (which is an extensional definition of the situation) Maxwell introduced field theory, which from its equations we may deduce the shapes of the lines of force, and the conditions obtaining at each point in the (x, y) field in which the magnet sits. The field strength at any point may be found, it decreases as we move away from the magnet. This static experiment reveals the resultant of forces, but does not give us a picture of the forces which are deducible from the field theory. To get an intuitive demonstration of the forces, take a second much smaller magnet and place it in axial alignment pole to pole (N to N, or S to S) at a distance z_x, z_y from the end of the original magnet, and let go. The small magnet then slides away and comes to rest at some distance away from its starting point. Again its behaviour is a consequence of the resultant of forces acting on it.

What has all this got to do with psychology ? The question is, do all these electromagnetic phenomena constitute a metaphor for phenomenological fields in vision, or for social fields in small groups[2] ? The analogy depends upon how tightly we interpret the metaphor, as simile, as analogue, or as formal isomorphism. Köhler (1920) wished to push the analogy home, in continuous space,

---

[1] Readers who are physicists will find this paragraph can surely be skipped.

[2] Lewins' social psychological excursions along this path are not considered here, apart from acknowledging that they exist (see Mey, 1972).

seeking mappings between electrophysiological fields in the cortex and fields in phenomenology (Petermann, 1932). Our new capacity readily to simulate the dynamics of connected networks has reopened these questions, but now in lattices as opposed to continuous fields. The sense in which the pixels of the lattice are connected, to some degree in a neighbourhood, is expressed in terms of the strengths, or transition probabilities, on discrete links between nodes. In this sense the mathematics of graph theory may take over from field theory to perform some explicatory tasks.

In fact there are two questions to be addressed; the first is what operations can be performed, if any, on the phenomenological field to demonstrate that activity at any point in the field has a vector representation, and the second question is to what extent networks (which are lattices connected in a predefined fashion) can simulate the field properties of the phenomenology. There are in fact at least three metatheoretical ways of addressing the second question, according to how reductionist one wishes to be. Within the strong Gestalt tradition, going back to Mach, which survives in psychophysiological models such as those created by Grossberg and Mingolla (1985), then the neurological concepts of mixed excitatory and inhibitory networks as substrate are used as isomorphisms with the resultant visual phenomenology. This is done for diffusion, including Mach bands as a special sort of diffusion, for brightness contrast and induction, and so on. If one wishes to simulate at the level of system outputs without regard to substrate mechanisms, and to preserve something of the field idea, the computer nets encoded in programs, perhaps series-parallel, will often serve, and are tractable for AI purposes. If one wishes to write a formal algebra for the phenomenology, using elementary local processes derived from other psychophysical situations not extended into fields, then lattice-connected psychophysics may be illuminating. In this last case the level of description, and perhaps explanation in the sense of restricted reduction to local processes at a lower level of complexity, does not necessarily involve the physiological substrate as an explicandum in the plot, but what is postulated has still, in principle, to be something that the brain could do.

Here it is necessary to tease out distinctions between the explicit psychophysiological reductionism of theorists like Grossberg or Marr, the looser attempts at isomorphisms of Köhler, the artificial intelligence neural networks of parallel distributed processing (PDP), and the nonlinear psychophysics (NPD) drawn on here as a starting point. There are differences of emphasis and intent, and indeed in the very idea of what constitutes an explanation of behaviour. The NPD approach makes fewer assumptions than some alternatives, and the assumptions are ones already employed in non-lattice nonlinear $n\Gamma$ theory. Explanation, in any case, should imply a systematic and economical account, not always one in which the current incomplete account of brain physiology is one with which the psychophysics has to be reconciled in every detail[3].

Potential, Vector and Force Representations

A comment is necessary at this point. Some of the Gestalt theories had been so strongly influenced by the notion that an electrical field was the appropriate precedent for modelling perception, that physicalist ideas of forces and vectors were imported verbally as descriptions of phenomenology. The local spatial distortions of rectilinear figures, which are a common result (physically straight lines become apparently curved) are sometimes claimed to be the consequence of forces or vectors inducing parts of the field to move, as though on an elastic and thus distortable surface.

In the treatment here, there are no continuous lines as inputs, only dot patterns in arrays, and no use is made directly of the ideas of vectors, or of forces. But mathematically it is possible, given the created continuous bivariate output surfaces in $Y_{xy}(\mathrm{Re})$ which are made with arbitrary assumptions about the settings of the hidden parameters of the bivariate spline (or thin shell spline surface as it is sometimes called), to create a second contour map in which at all points the new contours are now orthogonal to the original $Y(\mathrm{Re})$ contours as shown in the figures presented. No use is made in this treatment of the associated $Y(\mathrm{Im})$ surfaces which also exist at each Γ trajectory iteration and cascade step output, This second orthogonal surface, which in effect maps the steepest gradient in direction and magnitude at each location, is a picture of a vector surface. The forces which writers like to invoke are not in fact field properties *per se* but are output properties proportional to the vectors. Further, just as geographers like to create maps in which reproduced local areas are proportional not to square kilometres in the real world, but to populations within those areas, which in a sense are topographical but not topological distortions, so one could, given a vector or force map change its shape so that it becomes everywhere equipotential; in this case all gradients become equal and distances are stretched or squashed. Where there were originally no contours (flat local regions) nothing is changed, but steep hillsides flatten by locally

---

[3] The idea that we can seek psychophysical explanations across situations, but only within the level of the psychological data, is that pursued by Ekman and coworkers at Stockholm in their scaling of judgment phenomena.

stretching space. Lines which were straight originally would now become bent, if originally they were in the neighbourhoods of other lines.

Networked Nonlinear Psychophysics

An $(n \times n)\Gamma^k$ process is a bivariate form of nonlinear psychophysical dynamics, in which an $(m \times m), m >> n$, lattice of Γ trajectories, one at each pixel, evolves through time irreversibly. The term n defines the side of the square neigbourhood of each pixel with which it interacts, and only directly interacts. The form of the cross-coupling between each pixel and its set of $n^2 - 1$ neighbours is defined by the $n\Gamma$ **Case 2** dynamics (see chapter 1). The process is also cascaded, k being the number of cascades, the whole lattice is thus repeatedly fed back into itself; each cascade step is a set of $(m \times m)$ $n\Gamma$ trajectories in parallel, running for η recursions in the complex variable Y. The dynamics of the process thus run through a total time of ηk recursions, In a lattice using $n\Gamma$ **Case 2** cross-connections, but only using adjacent cross-coupling via λ, a pixel at (x, y) is connected simultaneously in the 3×3 case, for a rectilinear lattice[4], to each of its eight immediate neighbours, but they are not, in that iteration centered at (x, y), connected to each other. The computations must proceed sequentially by scanning across the lattice, $(x, y) \in m \times m$, and storing results until a full coverage on $m \times m$ (with lattice marginal effects ignored) has been made to input to the next cascade step. This induces a restricted pattern of diffusion in a local neighbourhood; the evolution of the trajectory of the pixel at (x, y) is then dominated by its strongest neighbour in that particular cascade step. So after η iterations of the Γ trajectory (within that cascade stage) the $Y_\eta|_{(k,xy)}$ will take the value arising from the λ coupling, which induces a minimum e equivalent. It follows qualitatively and quantitatively, that the pattern generated in each η trajectory of the total $m \times m$ lattice centered in turn on (x, y) will depend upon the initial difference between $a_{k,xy}$ and the a values of its neighbours. If a_0 is the background excitation initially imposed over the whole $m \times m$ $(m >> n)$ lattice, then any figure is a bounded set of local flaws a_f in the lattice, and $m \times m$ exhausts the region of the visual field being modelled, being itself initially partitioned into two discrete subsets coded by a_0 and a_f.

Thus imposing an image, in the form of a set of local flaws, on the pixels of an $(n \times n)\Gamma$ lattice involves making a number of parameter settings, the

---

[4] A hexagonal lattice is necessary if connection to fewer than eight neighbours is desired, six being then minimal.

interplay of which appears to be crucial to the outcome. What appears to be a local diffusion of images is associated with optimum values of e, that is of λ/a, in an $(n \times n)\Gamma$ lattice.

There are two processes to consider in $(n \times n)\Gamma$ dynamics; the interaction over the $m \times m$ lattice which is simultaneous (but which is stored in computations sequentially) and the cascading (of step k to $k+1$) where a complete lattice of local $Y_{k,xy}$, $xy \in m \times m$ values is mapped one-to-one onto a complete lattice of $a_{k+1,xy}$ values to begin a new lattice of Γ recursions. The properties of some examples of $(n \times n)\Gamma$ are studied by Gregson (1992a, chapter 11) with illustrations.

The $(n \times n)\Gamma$ structure within one cascade stage k is a locally connected lattice (LCL), which can be regarded as a state in a special sort of cellular automaton (Langton, 1986). The series of cascade steps, that together constitute the cascade itself, is an irreversible process in time whose evolution, pixel by pixel and step by step, is orthogonal to the plane of the lattice. The reader might like to visualise it as rising out of the page towards the eye of the viewer. To generate patterns which might resemble Mach bands, and some associated distortional phenomena in visual illusions, we explore for ilustration four variants (W1 through W4) of the generic $n \times n\Gamma^k$, using 3×3 and 5×5 neighbourhoods with local coupling through λ, and considering two forms of the $Y_k \mapsto a_{k+1}$ cascade recursion:

W1: 3×3 local interaction, meaning that each pixel interacts directly only with its eight immediate neighbours in a rectilinear lattice;

the cascade recursion equations from step k to $k+1$ are

$$\forall\ xy;\ a_{xy,k+1} = a_o + c_1(Y_{xy,k}(Re)_\eta - minY_k(Re)_\eta) \qquad [5.1.1]$$

$$Y_{xy,k+1}(Im)_0 = Y_{xy,k}(Im)_\eta \qquad [5.1.2]$$

where

$$c_1 = \frac{(a_f - a_0)}{(maxY_k(Re) - minY_k(Re))} \qquad [5.1.3]$$

The $minY_{k,\eta}$ term is based on the minimum generated over the whole $m \times m$ lattice at k.

W2: As for [5.1.1], [5.1.2], but interacting with 24 adjacent pixels in a 5×5 rectilinear neighbourhood.

W3: 3×3 local interaction;

the cascade recursion equations, dropping the xy suffices for simplicity, become

$$a_{k+1} = a_o + c_1(Y_k(Re)_\eta - minY_k(Re)_\eta)^{c_2} \qquad [5.3.1]$$

$$Y_{k+1}(Im)_0 = Y_k(Im)_\eta \qquad [5.3.2]$$

the purpose of c_1, a real positive scalar in [5.1.1],[5.3.1], is to rescale, and c_2 in [5.3.1] is to produce a partial correction for the Γ compression of $a \mapsto Y$ images. The setting of these constants has to be adjusted so that $2 \leq a_{k,0} \leq max(a_f)$.

W4: As for [5.3.1], [5.3.2], but interacting with 24 adjacent pixels in a 5×5 neighbourhood.

The common equation for a pixel interacting with its 3×3 neighbours is given (Gregson, 1992a, p. 226, eqn. 11.6) as

$$e_{xy,k+1} = \lambda \cdot (max(a_{x-1,y-1}, a_{x-1,y}, a_{x-1,y+1},$$
$$a_{x,y-1}, ---, a_{x,y+1},$$
$$a_{x+1,y-1}, a_{x+1,y}, a_{x+1,y+1}))^{-1} \qquad [5.4]$$

and this extends simply to the 5×5 case. Only **W4** is used to generate figures in this chapter. The superposed figures are a constant set of values a_f over a bounded subset of the pixels. These subsets are the analogues of the stimulus figures in some visual illusions depending only on outline drawings. No solid regions, or boundary contours between solid and empty regions, are considered here.

The outputs of the four variants can differ in their asymptotic resultant configurations with increasing k, and in their resemblance to psychological perceptual data. **W1** through **W4** can be treated as a 2×2 design; the dependent variables are $m \times m$ contour maps contingent upon fixed subset input a_f, the independent variables are some of $a_0, e_0, Y_o, \eta, k, \lambda, c_1, c_2$ and a_f.

Obviously all variants are identifiable deterministic processes in real time, time being represented by η, k and hence predictions might be derived concerning the consequences of eliciting visual judgments under paced or tachistoscopic presentation. The cascade recursion equations from step k to $k+1$ are **not** the same as the restabilisation equations of the previous chapter; there is no restabilisation within the cascades **W** as defined here; but instead a univariate recurrent remapping from Y(Re) to a.

Line Figures as Local Inputs

The extension of these ideas into the representation of geometrical figures is done by putting the line figures onto some of the pixels of a lattice; two a values, a_0 for the background (initially uniform) field level, and a_f for those pixels which constitute the input figure, are needed initially. Usually one would want to put $a_f >> a_0$, but this is not a necessary constraint. The case where $a_f > a_0$ is called a *positive image*, and where $a_f < a_0$ is a *negative image*. If it were desired to represent figures with different levels of figure-ground contrast within the same field, then a range of different a_f values would be used. A single line in a figure is represented by a single array, vertical, horizontal or diagonal, of pixels.

Resemblances to Cellular Automata

The theoretical interplay between chaotic dynamics and the processes within cellular automata has been explored in some detail in physics (Wolfram, 1984, Barnsdorff-Nielsen, Jensen and Kendall, 1993) but requires re-examination within nonlinear psychophysics, if only because the mathematical core is different from the attractors used in physics. If an $(n \times n)\Gamma$ lattice is used as a map of the phenomenological visual field, the various Γ trajectory parameters have to be set *a priori*, and the consequences of their settings then explored by extensive simulation. Most of the results are not accessible to intuition in advance, which is characteristic of nonlinear dynamics.

There is a family resemblance, which is misleadingly similar in some respects, between the sort of fields to be used here, and cellular automata (Wolfram, 1986). In both cases points interact with their neighbourhoods, in a recursion through time and space, and the patterns created are generated quite critically by values initially given to points, and the size of the neighbourhoods used in local recursions within a cycle. In both cases it is possible, with very small parameter changes, to obtain various distinct types of dynamic evolution. Fields may completely homogenise and effectively vanish, limit cycling may emerge, chaos may supervene, and some hybrid forms difficult to categorise or to identify may be found. Wolfram (1986) suggested this four-fold classification for cellular automata, and something analogous seems to obtain for nonlinear psychophysical fields. The so-called *edge of chaos* is a place to look for interesting and unpredictable properties, where signals can reorganise themselves into sustained stable but complex structures, and float back and forth into and out of chaos; it is not critical but it is quite possible as a heuristic to see what is being done here in theoretical psychology as a sort of complexity theory (Waldrop, 1992). Categorising dynamics under popular titles may give a feeling of familiarity, but ultimately comparison of simulated with real data is essential. It is important to note that some of the patterns generated in cellular processes cánnot be represented analytically, using symbolic dynamics, in any way simpler than the patterns themselves; that is, the shortest symbol string that will generate the pattern is as long as the pattern itself.

Field theories can themselves be broadly divided into two sorts, and each has had its offspring in psychology. The distinction beteween process and outcome models drawn in chapter 4 is germane here. From a description of perceived fields, as in Gestalt psychology or in the cataloging of types of illusions (Rausch, 1966, Robinson, 1972) only outcomes are provided, whereas in both cellular automata, and in the evolution of $(n \times n)\Gamma$ fields there is a possibility of tracing intermediate steps as well as predicting final outcomes at any point where the evolution is arrested. The non-Euclidean algebra of perceptual spaces can be modelled by Lie algebras (Dodwell and Caelli, 1993) as an outcome description, but the evolution in real time to create such spaces is then implicit and not explicit.

Evolution of the Process

The $(n \times n)\Gamma^k$ process may be examined at any intermediate stage, at each of its possible ηk steps; such intermediate plots of $m \times m$ pixels are made up of a or $Y(\mathrm{Re})$ values at the stage examined.

Because the cascade goes through a sequence in which $(n \times n)\Gamma$ and equations [5.4 - 5.6] alternate, in each cascade step, the two types of lattice plot and the contour approximations to a field are readily derived. However, none of these necessarily correspond directly to a prediction of an outcome of the dynamics; what is also needed is some representation which corresponds not to an unobservable internal process, in real time, but to the reported phenomenology of the perceptions associated with the input figure in its given a_f, a_0 values. To attack this the mean entropy H and cross-entropy B maps can also be constructed. Quite simply, if $1, ..., x, ..., m,\ 1, ..., y, ..., m,\ 1, ..., t, ...k$, then the $m \times m$ $Y_{xy,t}$ values at each of the cascades are affine renormalised and then replaced by the $\theta_{xy,t}$, $0 \leq \theta \leq 1$ so that averaging over the k cascade series gives

$$H_{xy} = \frac{-1}{k} \sum_{t=1}^{k} \theta_{xy,t} log(\theta_{xy,t}) \qquad [5.5]$$

This represents the persistent field invariances within the evolution of the process. What usually happens is that the region originally associated with a_f is strongly emphasised, and second-order turbulences are sometimes revealed around it. This is like emphasising figure-ground contrast.

That is, plotting the series of $Y_{xy,t}(\mathrm{Re})$ $m \times m$ intermediate process lattices (and their contour maps) for a given a_f, a_0 input pattern and levels of contrast would display much more turbulence than the results of [5.5]; it is useful for displaying at what value of ηk a dynamic process converges to stability, fluctuates[5], become chaotic, or explodes, but it has no necessary one-to-one correspondence with ultimate output. In contrast, [5.5] is a sort of averaged mapping with local entropy weighting.

A central problem of psychophysics is to evaluate the correspondences, or lack thereof, between stimulus and response; to discover if the invariances in the process represented by [5.5] do reproduce the stimulus input pattern, in short to see if perception is veridical. The cross-entropy (Boltzmann entropy or Kullback-Leibler information measure) is a representation of such correspondences. Put u_{xy}, $0 \leq u \leq 1$ as the normalised a_{xy} matrix and then

$$B_{xy} = \frac{-1}{k} \sum_{t=1}^{k} \theta_{xy,t} log\left(\frac{\theta_{xy,t}}{u_{xy}}\right) \qquad [5.6]$$

This B goes to zero in regions where an input image is transmitted undistorted. Contour maps of B will exhibit sharp gradients if a_f regions are transmitted differently from their immediate surrounds, and if strong persistent Mach bands are generated. In the graphs here [5.6] is mapped where noted; one graph for each λ value over the range 0.83, 1.41.

It is important to note that the total length in time of the process, ηk, can be produced in many different ways, here the situation with $\eta = 5$, $k = 19$ is used first to explore the consequences of treating λ as the only system variable. What might be called a converse representation (using *converse* very loosely) is made with $\eta = 19$, $k = 5$. This runs over the same length but may have very different internal dynamics, it may even be impossible because it explodes or dies.

[5] If the process does not show fluctuations between if not within cascade steps, then it is psychophysically very implausible. This comment relates to the observed oscillations in response latency distributions (Gregson, 1992a, chapter 3, 1993a, Dehaene, 1993) which are identifiable but not strictly phase-locked.

Arrow Tail Figure Transformations

The basic figure used in this chapter is very simple, a short horizontal line splits into two fins at 45°; this is like one end of the ubiquitous Müller-Lyer figure with extending arrow heads. Discussion of other figures which also have a very long history, and detailed results in experimental psychology, is treated in the next chapter. The interest in this figure arises because the concave region within the fins can be filled, or other regions on the axis of symmetry of the figure enhanced, during the $(n \times n)\Gamma^k$ process evolution. If this evolution is postulated as a real time mechanism for illusion generation then the a_f, a_0 values are stimulus dependent, but the λ is an internal variable of the observer and not accessible to experimenter control. At the same time, ηk and λ should be regarded as unstable system parameters, and the generation of variance or uncertainty in the responses recorded should in part be correlated with such internal parameter instability.

Simulations

Comments on the Figures 5.2 through 5.12 (which all show cross-entropies) are keyed by their λ values, all other parameters in [5.4 - 5.6] are fixed over the series: $\eta = 5, k = 19, c_1 = 3.0, c_2 = 1.5, Y_0 = (0.5, 10^{-4}), a_f = 4.8, a_0 = 2.8$

Fig.5.1 The basic input pattern of the fins on an arrow, ecoded as a_f pixels in an a_0 ground.

Fig.5.2 $\lambda = 0.83 to 0.93$: Increasing λ makes no difference so far, in fact the system is λ-robust for a wide range of values up to around $\lambda = 1.0$ This is compatible with results in other psychophysical contexts.

A value of $\lambda = 1.0$ is equivalent here, initially for $k = 0$, as $e_f = 1/2.8 = .357, e_0 = 1/4.8 = .207$, which should give stronger representation for the figure pixels.

Fig.5.3 $\lambda = 1.03$ Here things become interesting; the input figure is present ($B \simeq 0.0$) but the region within the acute angle made by the fins is now created as a fluctuating unstable zone. The diffusion bands are still present, and some small asymmetry appears at the ends of the fins, indicating that some one pixel has gone past a Y bifurcation, but not necessarily onto a strange attractor.

Fig.5.4 $\lambda = 1.04$: This is a slightly more stable form of the previous figure, it is a transition to the next configuration.

Fig.5.5 $\lambda = 1.055$: The transition in form induced by increasing from $\lambda = 1.03$ to $\lambda = 1.05$ is rapid.

This and the next few graphs are based on quite small increments of λ;

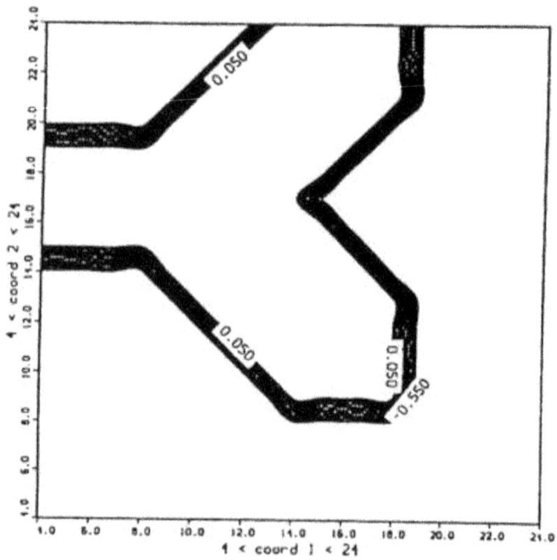

Figures 5.1 and 5.2

the dynamics change qualitatively quickly and a cautious interpretation is in order. The figure 'tidies itself up' back into symmetry and the region within the angle is distinctive but less extreme in level.

Fig.5.6 $\lambda = 1.08$: Another small increment in cross-coupling resistance; the figure is strengthened and the diffusion pattern has smaller B gradients, again with some slight asymmetry.

Fig.5.7 $\lambda = 1.13$: This figure is again tidy, and resembles the situation around $\lambda = 1.05$ but with less activity within the acute angle.

Fig.5.8 $\lambda = 1.23$: The diffusion pattern has vanished, the arrow figure is even more contrasted with its surround, and the angle filling is marked.

Fig.5.9 $\lambda = 1.33$: From here onwards it is necessary to proceed cautiously to avoid exploding the a, Y iterations.

The asymmetry in B is marked and sprawling, what is of interest in the region spanning the horizontal shaft on the left of the figure. Angle filling persists.

Fig.5.10 $\lambda = 1.35$: An increase in λ of .02 is sufficent to alter markedly the diffusion pattern; now the pattern on the left goes away, but a new remote region on the axis of symmetry emerges on the far right, at some distance from the figure.

Fig.5.11 $\lambda = 1.39$: This graph is almost symmetrical in the neighbourhood of the input figure, the region on the axis of symmetry has moved to the left and into the acute angle, and the angle filling is again marked.

Fig.5.12 $\lambda = 1.41$: This is about the upper limit before the $\{a_f, a_0, \lambda\}$ parameter set chosen will induce explosions in Y. Traces of previous effects and asymmetry reappear markedly, the angle filling is much less in evidence.

A tentative conclusion, but quite compatible with previous applications on nonlinear psychophysics with $n\Gamma$, is that the perceptual processes which arise in real time, to generate the Müller-Lyer illusion, would be maximally associated with λ values around 1.2, for the amount of figure-ground contrast used.

Variances of Cross-entropy in the Process Evolution

So far in the argument, and for all the figures, the average B_{xy} values over k cascades have been used as the descriptive statistics in order to characterise the information transformations arising. A mean is hardly a sufficient statistic, and for a very long cascaded series one might want to examine as well the time-series properties of the evolution of B, even though under many parameter settings

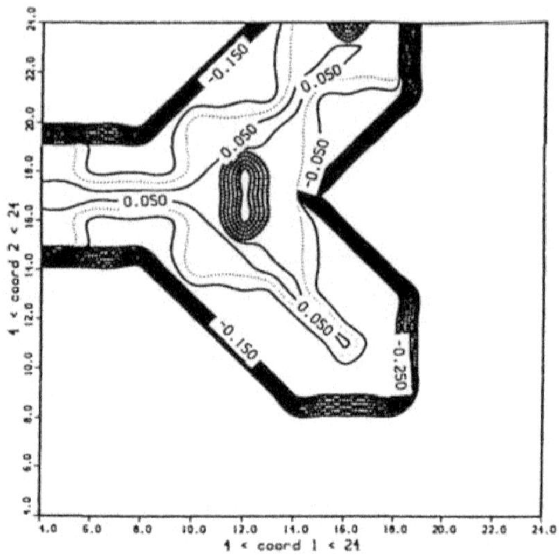

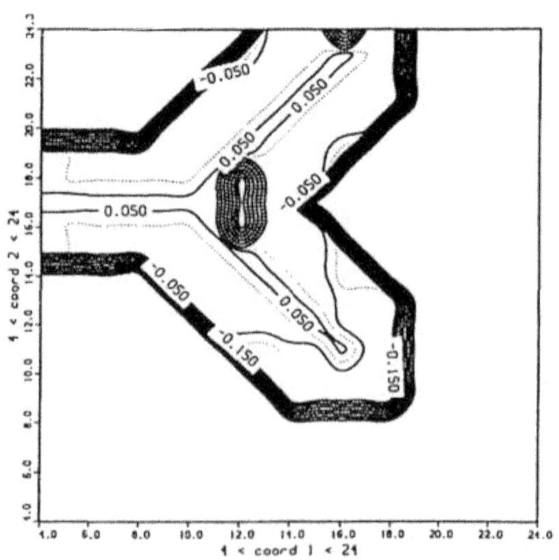

Figures 5.3 and 5.4

the cascade series will stabilise on a ganzfeld map and so is not stationary up to that point, when it becomes trivially stationary.

The variance of B over the k series can be interesting in that it points up the regions of instability, and hence of potential ambiguity in perceptual interpretation. Instability here is more than information in the H or B senses, it is a fluctuation in real time of B for given starting conditions. To illustrate the difference between mean B and $varB$ maps, some values of λ have been picked which look interesting, in Figures 5.13 through 5.17.

Parameter Settings for Figures 5.13 through 5.18

For all figures, $\eta = 5$, $k = 19$, $a_0 = 2.8, a_f = 4.8$.

Fig.5.13 $\lambda = 1.23$

Fig.5.14 $\lambda = 1.41$

Fig.5.15 $\lambda = 1.45$

Fig.5.16 $\lambda = 1.51$

Fig.5.17 $\lambda = 1.6501$

Fig.5.18 $1.23 < \lambda < 1.6501$

They should be compared with the previous corresponding figures, though the range of λ values is now 1.23 to 1.6501. The value of $\lambda = 1.6501$ is at the upper limit for non-exploding evolution of $(n \times n)\Gamma^k$ given the other parameter settings used throughout. Again, the concave region within the angle is usually filled, but there is more evidence of effects about the axis of symmetry than is hinted at by the mean B figures. The fins are carrying strong information as before; this is because they show flip-flop in Y values both within and between cascade steps.

A comparison with a conclusion of Gilden, Schmuckler and Clayton, (1993, p. 478) is pertinent. From an analysis of fractal boundary perception, in which mathematically fractals are by definition hierarchically ordered with self-similarity properties, they deduce that perception either cannot or does not utilise such self-similarity. They write "Perception, however, does not appear to operate in terms of hierarchies. Rather, its agenda is one of establishing a two-part division in which coherency and invariance are distinguished from noise and variability. This agenda is enforced in spite of the prevalence of structural alternatives". To see the parallel with what is happening here, consider the following conceptual extension of the situation explored with the series of fixed λ values in the cross-entropy variance plots.

As λ is the internal variable so far taken as constant, relax the process

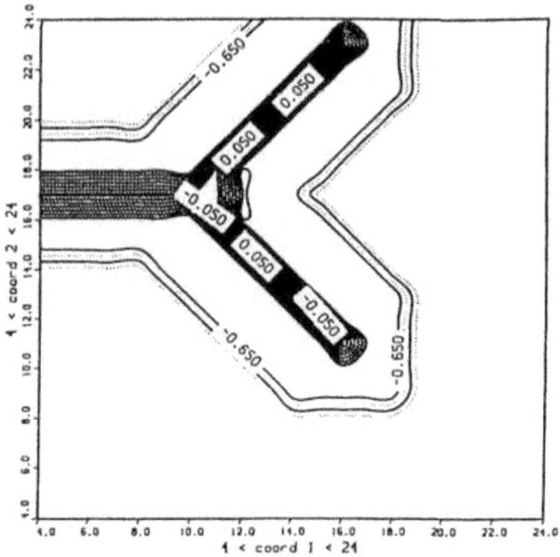

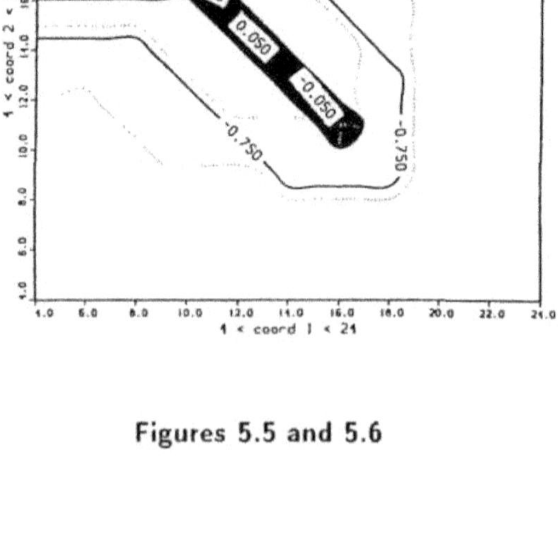

Figures 5.5 and 5.6

structure very slightly and allow λ to have a second-order variance about a mean within the near chaotic range already mapped. Then the plots will become a weighted compromise of the figures shown. This is a way of integrating the dynamics over time and the parameter space. In terms of summations, we now would have

$$V_{B,xy} = f\{const \cdot \sum_{\lambda min}^{\lambda max} \sum^{k} \sum^{\eta} B_{xy,\eta k \lambda}\} \qquad [5.7]$$

and this V_B would approximate to white noise over the whole $m \times m$ field, *except where the figure lies.* This can be shown by direct calculation; the contour map *of the ground* is now blank except for one weak diffusion contour, and this is shown in Figure 5.18.

There is some point in emphasising the distinction used here between a process with superimposed stochastic noise, and a nonlinear deterministic process with no stochastic (random) components, but with second-order instability in its parameters during process evolution in real time. Algebraically the two are quite different, for example

$$y = a + bx + \epsilon, \quad \epsilon \sim N(0, \sigma_\epsilon) \qquad [5.8]$$

is the familiar linear model with random error, which shows itself in y in the output over replications, but here we have

$$Y = \Gamma|_{a,\lambda}, \quad \lambda \sim W(mean\lambda, \sigma_\lambda) \qquad [5.9]$$

where in [5.9] W is some bounded distribution with first and second moments, lying entirely inside the range of the near chaotic dynamics. The distribution over time is now inside a single replication of the ηk duration process, not over a series of its replications.

This means the stability of the dynamics over a field on which an input line figure is imposed is effectively partitioned and the ground of the figure becomes almost white noise instead of completely uniformly flat. This is a sort of induction by the figure (into its surrounding ground), which does not destroy itself in the process though it may become blurred and distorted. At the same time, buried in the now noisy field are the second-order effects that emphasise symmetry, fill angles, and induce something like Mach bands. These second-order phenomena need never come into conscious experience, if by conscious experience we mean the resultant of processes such as [5.7], when the system

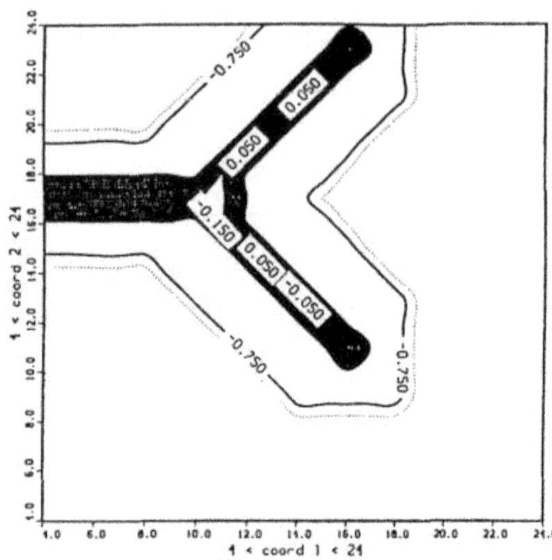

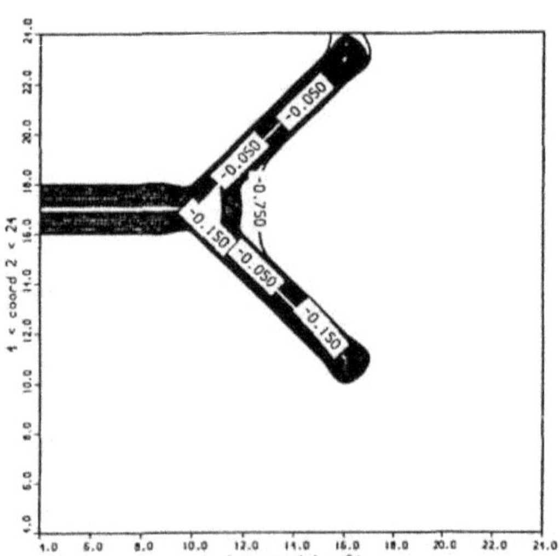

Figures 5.7 and 5.8

is operating in the near chaotic region of its dynamics. But their underlying presence could affect the relative speed and efficiency with which such things as symmetry judgments are made (see Gregson, 1992a, chapter 11 for a closely related discussion on symmetry detection); the presence of the figure primes the system to do other things related to the shape of the figure, if a field theory representation is valid.

Quantification of Illusion Extents

Some studies report the mean extent of the Müller-Lyer illusion under various conditions, usually as a percentage error in line length estimation. This raises two questions; is the mean a meaningful statistic, and how does one proceed from the plots of contours of $Y(\mathrm{Re})\eta, k$ to get an estimation of the perceived length of the shaft in the figure, given the angle filling effects within the evolution of the process?

The questions are not necessarily valid ones to put, not simply because the mean is hardly a sufficient statistic in such complicated dynamics, but because the sample mean may not exist as an estimate of the first moment, but only as a local sample description. In such circumstances, the reports of differences between findings in experiments have a dubious status; there need not be any true single answer to the question as to how big an illusion is. Box and Tiao (1965) showed that some time series can exist but have no first moment, and essentially the trajectories of the Y_{xy} pixels in $(n \times n)\Gamma^k$ are complex time series of length ηk. As a consequence of the way the cascades are stepped, these Y_{xy} series are not continuous in their derivatives and not stationary, but they are completely definable and deterministic. After such a series is completed, its sample statistics are derivable, and their *relative variances* at different locations in the matrix can be computed. These, or the entropies, are predictors of relative uncertainty of different points over the field; it should be possible to relate the phenomenological evidence about uncertainty of regions near line segment junctions to the heteroscedascity of the process output.

For example, it is possible to track the evolution of specific pixels in the fin diagrams used here, over the total evolution of the process, and show that the amplitude and the quasi-periodicity of the trajectories differs between points on the line shaft and points within the concave angle region which becomes sometimes filled. This is considered in the next section.

Synchrony and Asynchrony in Trajectory Evolution

To understand in greater depth what is happening as the lattice evolves in

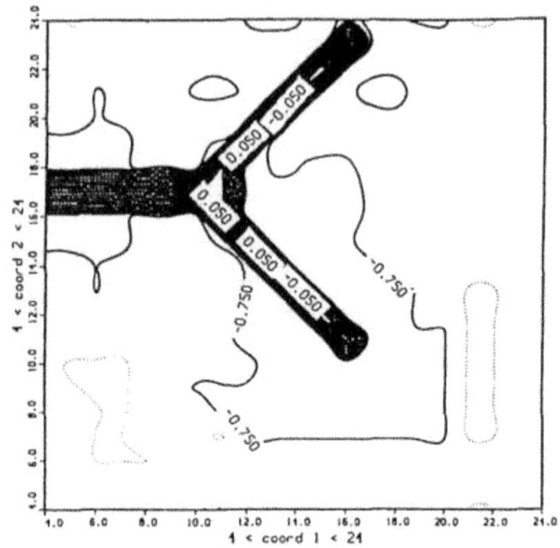

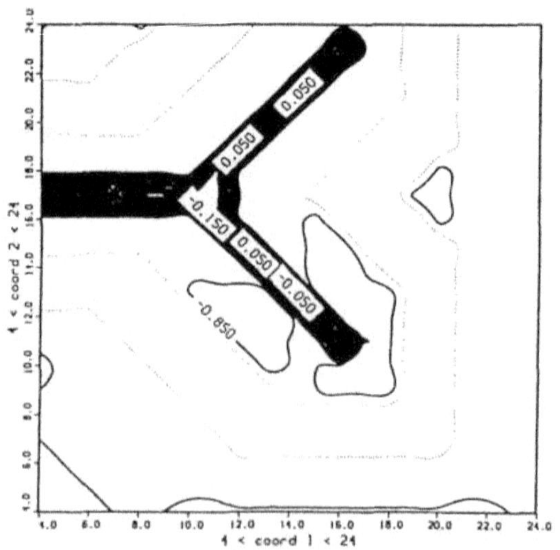

Figures 5.9 and 5.10

the real time span represented by ηk, the actual trajectories of $Y_{xy}(\text{Re})$ values, for some selected pixels, have to be examined in detail.

Parameter Settings for Figures 5.19 through 5.32

Fig.5.19	$\lambda = 0.90, \eta = 5, k = 19$
Fig.5.20	$\lambda = 0.90, \eta = 6, k = 16$
Fig.5.21	$\lambda = 0.90, \eta = 7, k = 14$
Fig.5.22	$\lambda = 0.90, \eta = 9, k = 11$
Fig.5.23	$\lambda = 0.85, \eta = 5, k = 19$
Fig.5.24	$\lambda = 1.10, \eta = 5, k = 19$
Fig.5.25	$\lambda = 1.10, \eta = 6, k = 16$
Fig.5.26	$\lambda = 1.10, \eta = 7, k = 14$
Fig.5.27	$\lambda = 1.10, \eta = 9, k = 11$
Fig.5.28	$\lambda = 1.35, \eta = 5, k = 19$
Fig.5.29	$\lambda = 1.10, \eta = 9, k = 21$
Fig.5.30	$\lambda = 1.35, \eta = 11, k = 15$
Fig.5.31	$\lambda = 1.4, \eta = 7, k = 28, c_2 = 1.0$
Fig.5.32	$\lambda = 1.387, \eta = 5, k = 39, c_2 = 1.5$

Two series of plots are used here; first a series (Figures 5.19 through 5.22) comparing three points, all lying on the symmetry axis of the fin figure, (7,17) on the shaft at the left, (12,17) inside the angle between the two oblique fins, and (16,17) out on the right in the ground. Thus the first point is in the figure, the second and third are in the ground, and the second is closely adjacent to the figure.

The parameters to be considered in plotting $Y(\text{Re})$ evolution are λ, η and k, for fixed a_0, a_f and using $(5\times5)\Gamma$ neighbourhoods. The values of $\lambda = 0.9, 1.10$ are picked for their contrasting dynamics. It is expedient first to show with a fixed λ that using η, k values of 5,19; 6,16; 7,17; and 9,11 makes relatively little difference to the most striking phenomena, that is, the contrast between the evolution of the pixel within the angle as compared with the other two points on the symmetry axis. The three series are not synchronic, and the $Y(\text{Re})$ values of the pixels in the angle die quickly if $\lambda \leq 0.9$, those on the ground rise and stabilise. The plots are all shifted on the y axis, by amounts shown in brackets on their labelling, so that details in form rather than absolute values are emphasised. The extent to which nonlinear trajectories are apparently periodic, as opposed to quasi-periodic or chaotic, varies with the choice of η, k. The products ηk have all been picked to give comparable series lengths between

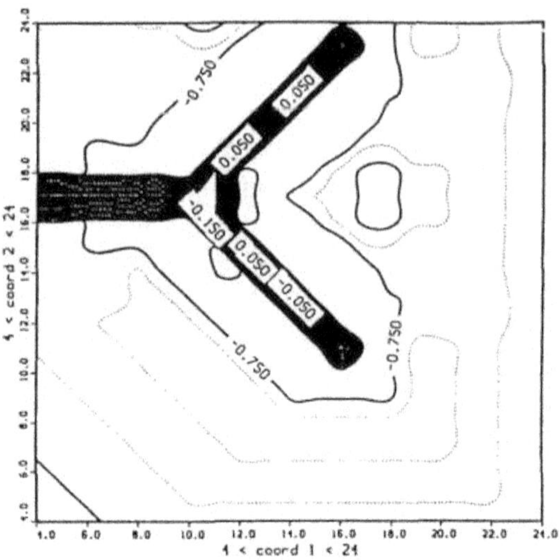

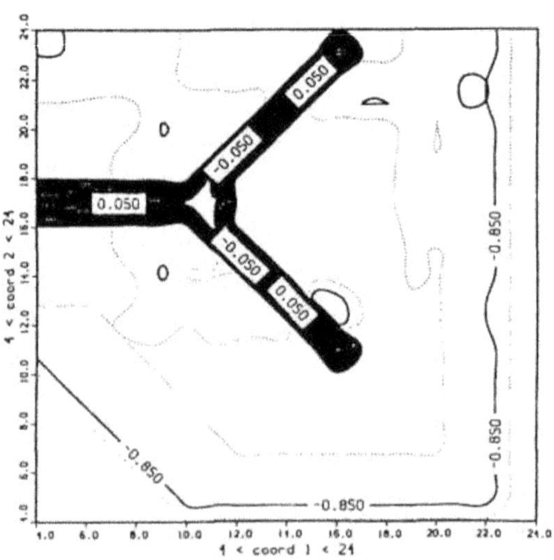

Figures 5.11 and 5.12

90 and 100 iterations in this and the next set of plots.

Using $\lambda = 1.10$ (Figures 5.24 through 5.27) is much more interesting in that it demonstrates the speed and extent of angle filling which plays a supposedly key role in illusion generation. Here again, the major contrasts in dynamics between the pixel in the angle and the other two pixels on the symmetry axis are robust (i.e. Andronov stable) with respect to the choice of η, k. The strong effect of angle filling is over and completed by 30 iterations, which mostly is less than three cascade steps. The larger η, the slower the effect runs to completion, indicating that the restabilisation of $Y \mapsto a$ between cascade steps is the critical mechanism. There is a tendency for the shaft and ground trajectories to evolve to near synchrony, but it should be emphasised that aperiodicity is the dominant pattern. For $\lambda = 1.10, \eta = 6$, the ground (initially a_0, Y_0) appears to run to periodicity, via a synchronic locking with the shaft pixels. It is clear that the dynamics within the local region susceptible to angle filling are quite asynchronous with the rest of the figure on the symmetry axis. The plots (Figures 5.23,5.28) of $\eta, k = 5, 19$ for $\lambda = 0.85$ and for $\lambda = 1.35$ are shown simply for extreme contrasts. This last case (Figure 5.28) is well into chaotic dynamics; trajectories are aperiodic (nonstationary in frequency) and the pixel in the angle is now much more unstable than the figure and ground points. The initial rise to fill the angle is rapid.

The last two of the plots (Figures 5.29, 5.30) again represents three different pixel trajectories, and again trajectories are shifted up or down on the graph by amounts indicated in brackets on the right, to exhibit asynchronous contrasts. The pixels used are now points on a fin, from the root near the angle to the tip.

All the plots up to this stage have used less than 100 iterations ($\eta k \leq 100$). A quite dramatic result is, however, revealed if instead three points all within the figure, but on the fins, are used, and the process is allowed to run for much longer. The points have the same initial (a_f) condition, and if there are no figure-ground or figure-internal interactions contingent upon the topology, they should remain with the same synchronous trajectories for given system parameter settings λ, η, k.

In fact for about 90 iterations this is precisely what happens. But then, as a consequence of the internal dynamics, and without any external inputs, there emerges asynchrony which for $\lambda = 1.10$ separates the fin root from its neighbours along the fin, and for $\lambda = 1.35$ dynamically differences the middle of the fin. At the same time the strong aperiodic synchrony of the root and end can persist when the middle becomes asynchronised. Once trajectories are

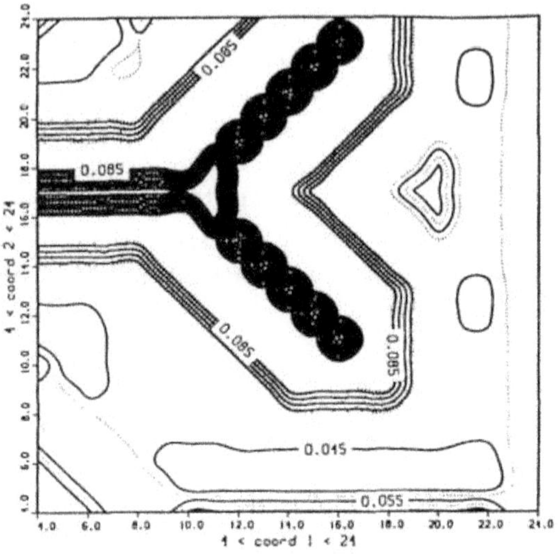

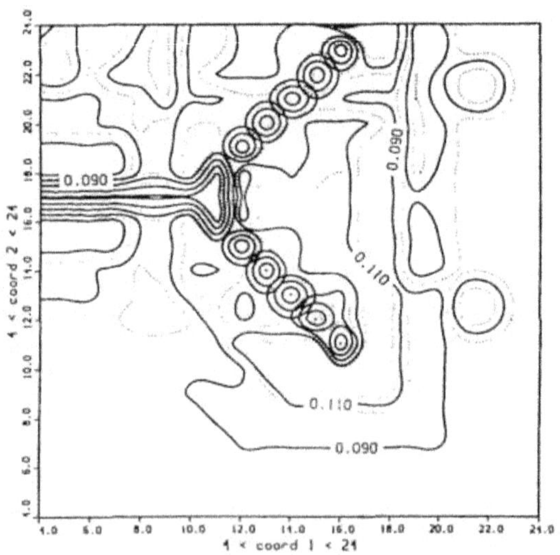

Figures 5.13 and 5.14

asynchronous, in this way, the probability of their being locked together again into synchrony is extremely low. The topologies of the neighbourhoods of the three fin points used to plot Figures 5.31, 5.32 are obviously not the same; the interest lies in the possibility that points at differerent distances along the fin from the line junction have quite different field dynamics; the length of the fin in pixel units must therefore come into the picture. The result of Pressey and Pressey (1992) that the apparent magnitude of the Müller-Lyer illusion is not simply related to fin length, but rises and falls as the fins lengthen, is compatible with this asynchrony. The abrupt changes in dynamics around $\eta k \simeq 90$ is almost certainly related to a jump across a bifurcation in the inbuilt Γ dynamics. But it is not the same sort of phenomenon as the "capture" of the trajectories for angle filling and ground which occurs initially in $(n \times n)\Gamma^k$ type **W4** ($c_2 = 1.5$ in [5.4]) which shows in Figures 5.24 through 5.27.

Indeed, instead of being a capture into asynchronous oscillations of comparable amplitudes, it is a complete shift from one pattern of synchronically coupled aperiodic trajectories to uncoupled aperiodicity of a different sort. The two examples shown are cases where one trajectory becomes asynchronous with respect to the others. It may therefore be asked, does there exist a region in the parameter space of the system in which uncoupling induces more than two disparate trajectories within pixel sets with the same starting values ? The answer is simply yes, over a diversity of parameter settings, using c_2, η, k and λ. But the local topology of pixels, due to the construction of their neighbourhoods, is not isomorphic with their given starting values which define the initial figure-ground relationships.

Putting Noise into the Ground

Adding noise to a system which is governed in its noise-free dynamics by a strange attractor does not have simple consequences, what happens depends crucially on the level of the noise (Kapitaniak, 1990). Noise is unpredictable and usually simulated by trajectories which have a very high fractal dimensionality, only its moments may be defined and possibly estimated from samples. In NPD by definition noise, in this sense, can only arise from inputs to the system, not from within the system.

For illustration, the a_0 values of the pixels in the ground part of the lattice were perturbed by random increments δa rectangularly distributed at second-order levels, a

$$0 \leq \delta a \leq c, \quad \text{is a system parameter} \qquad [5.10]$$

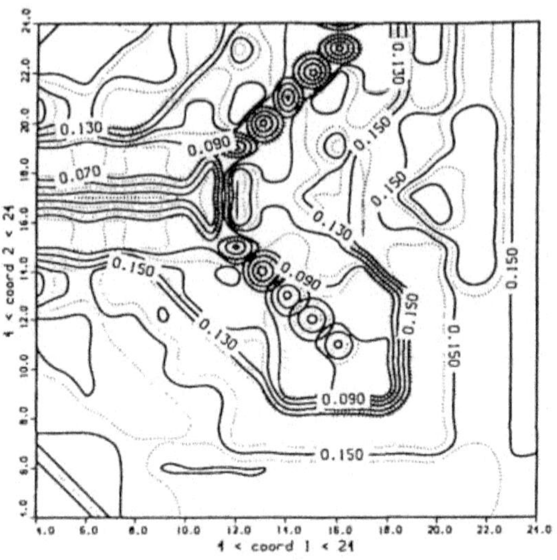

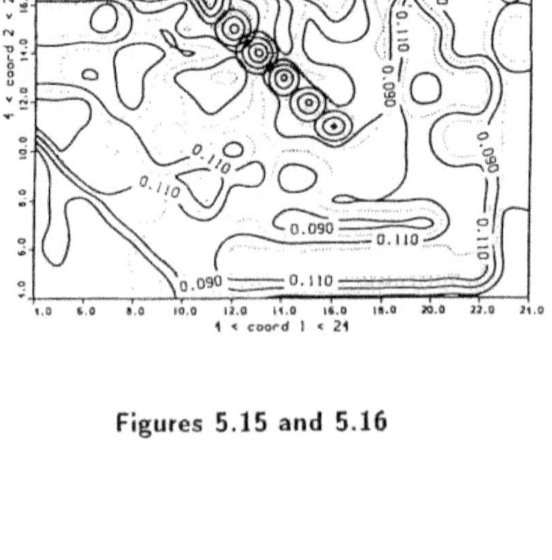

Figures 5.15 and 5.16

The a_f pixels in the figure were not initially altered; evolution of their values into heterogeneity or asynchrony is still due only to the $(n \times n)\Gamma^k$ evolution, for fixed η, λ. The situations explored up to this point have, by definition, been those with $c = 0$ in [5.10]; interest focusses now on what happens as c is very gradually increased. The mean δa is always $c/2$.

It only needs very small $c \simeq 10^{-3}$ to produce visible spotty perturbations in the ground, and to create a curiously turbulent region in the diffusion bands (Figure 5.33). But the figure, as it were, now resists the weak noise and induces an empty local region around itself, as though to repel degradation from the intruding noise. Weak noise actually helps to enhance the figure's bounds. Increasing to $c \simeq 5 \times 10^{-3}$ (Figure 5.34) is sufficient to encroach on the figure; it is then almost lost in a chaotic field unless one knows exactly where to look for the locus of the a_f pixels. This exploitation of, and capitulation to, ground noise is critically dependent on the λ, η, k parameter settings. It appears that the situation is very vulnerable and if edge-of-chaos dynamics are already present (not the $\lambda = 1.10$ used in Figures 5.33, 5.34) then the situation is degenerate; noise amd chaos at the same time can be just too much to handle.

Conclusions

The dynamics of $(n \times n)\Gamma^k$ are rich and intricate, particularly when the parameter space is explored over an area which spans values already used in previous psychophysical applications.

In a network context, phenomena that arise as a consequence of the bivariate plotting and the cascade mappings used here do constitute a surprising and even counter-intuitive diversity, resembling properties which had sometimes been anticipated in the purely verbal field theories of Gestalt psychologists. With the scale (that is, resolution) of lattice used here, which is better expressed as the compromise between neighbourhood span $(n \times n)$ and figure mapping (width of rectilinear regions set at a_f, here one pixel as minimal) it has been possible to induce figure-to-ground inductions of considerable intricacy. Extension to lattices with very fine scale[6] (perhaps nearer to some representations of a physiological substrate) is not necessary for the task of illustrating, in principle, what a field theory derived from, and only from, the minimal assumptions of NPD will generate. In any case, the 27×27 fields used here cover only a tiny fraction of what a total visual field would need in

[6] The resolution is in fact effectively enhanced by the continuity assumptions built into the bivariate spline fitting used to generate the contour maps.

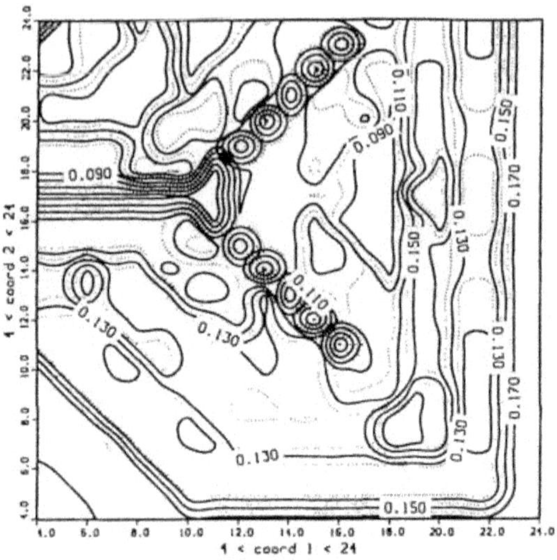

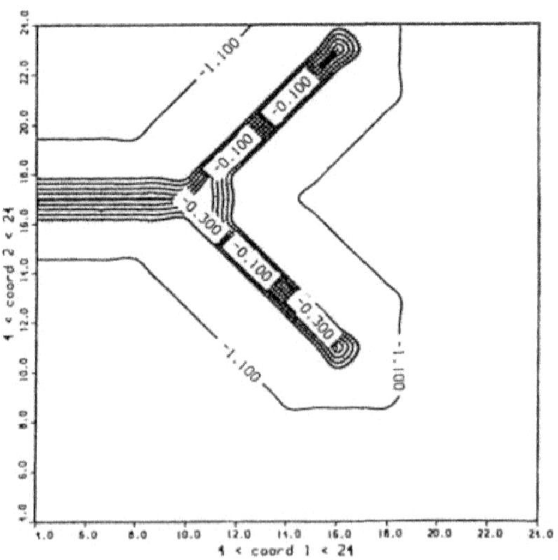

Figures 5.17 and 5.18

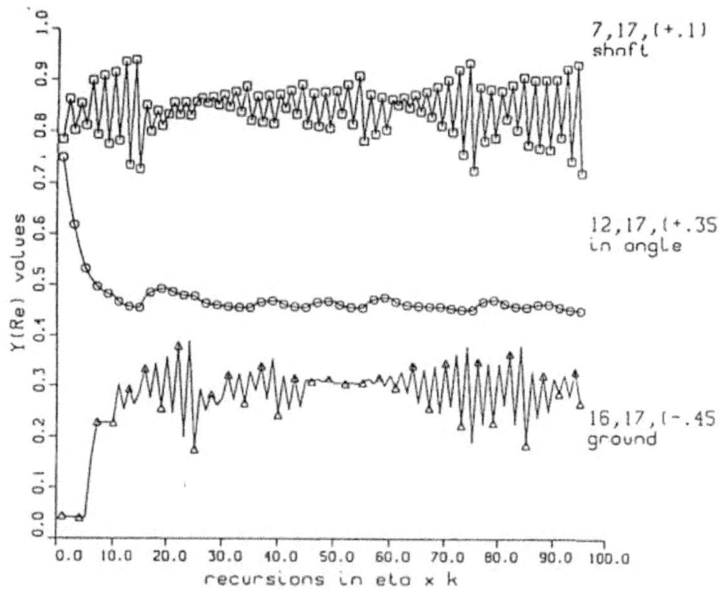

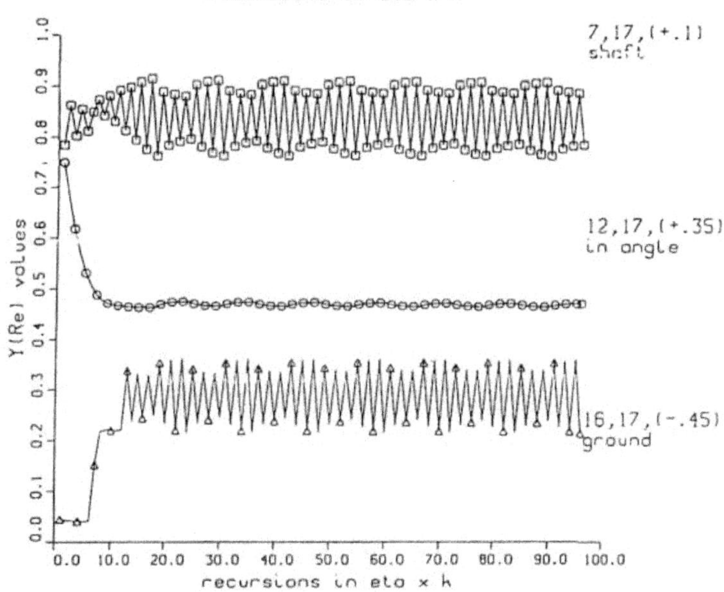

Figures 5.19 and 5.20

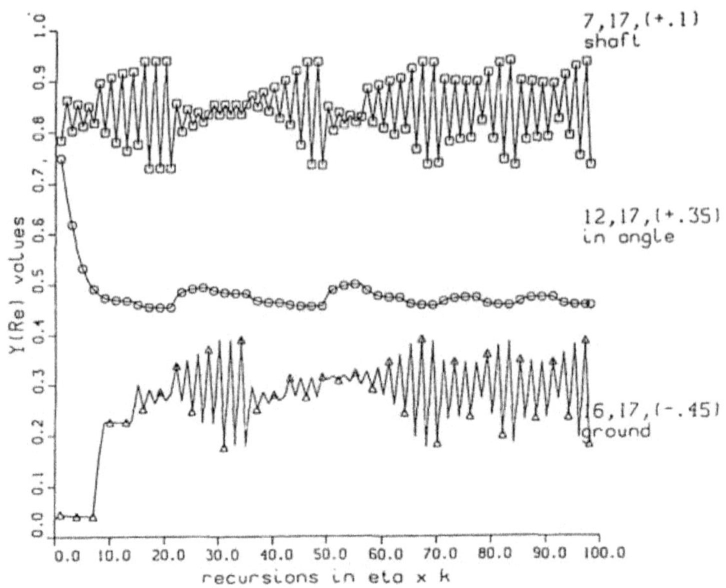

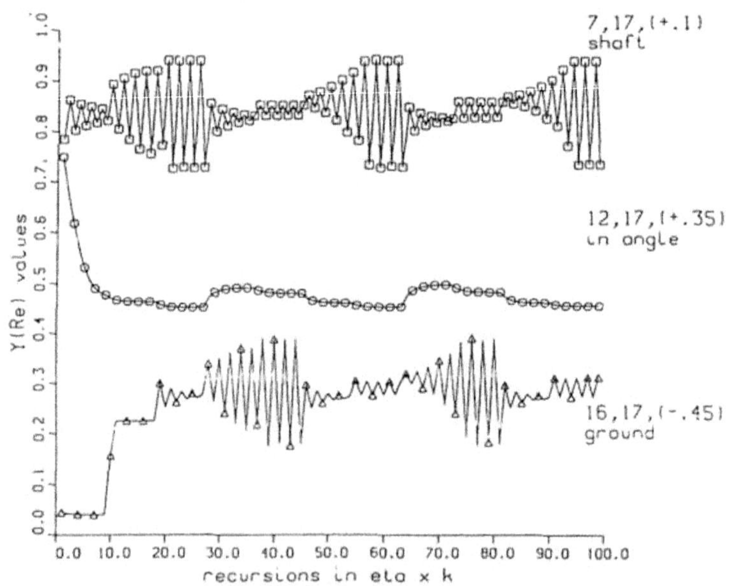

Figures 5.21 and 5.22

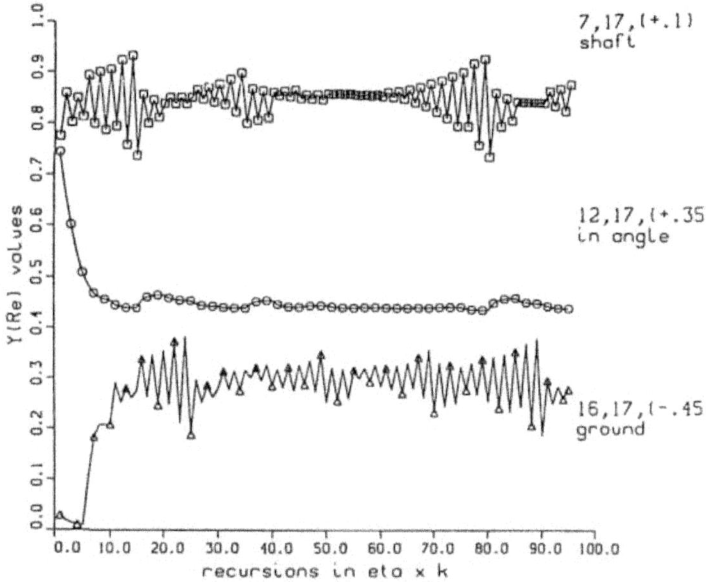

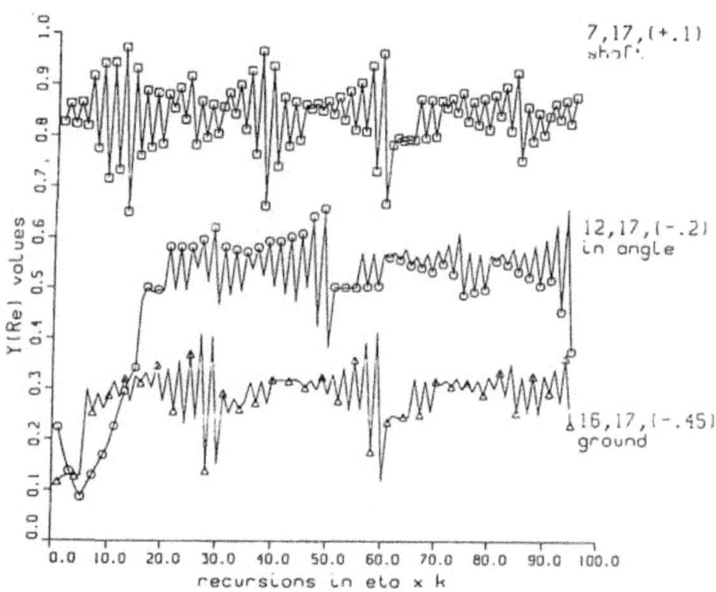

Figures 5.23 and 5.24

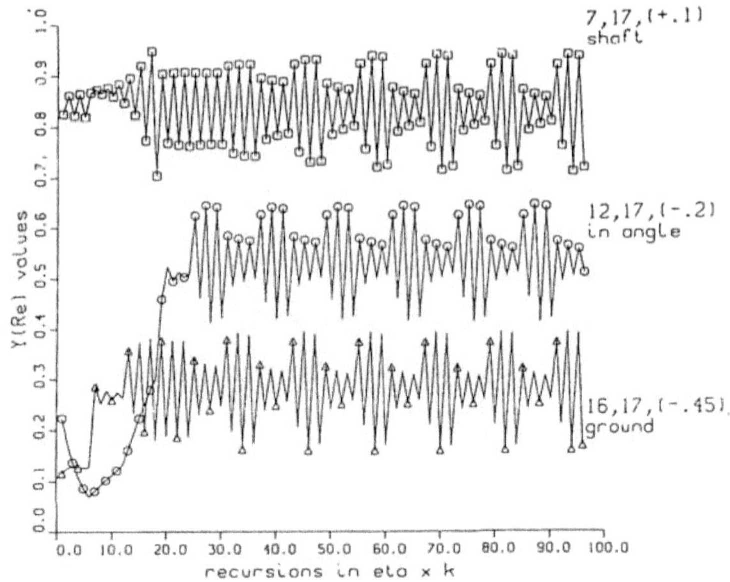

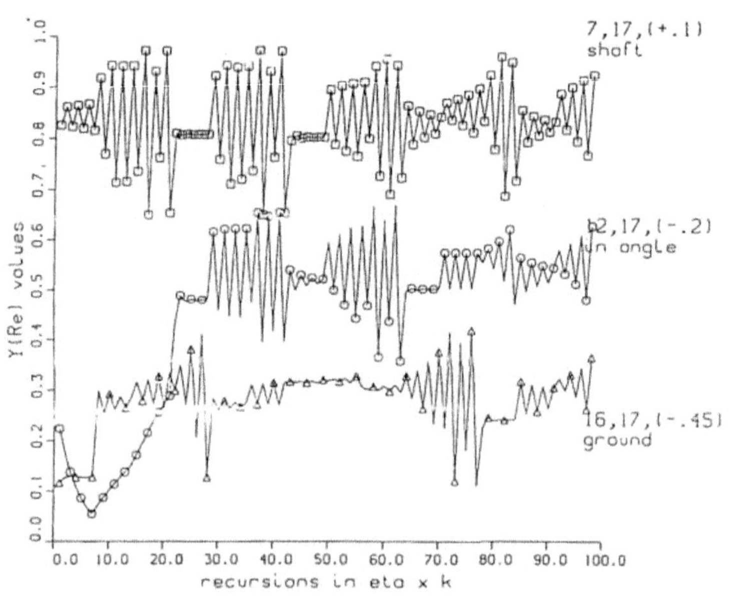

Figures 5.25 and 5.26

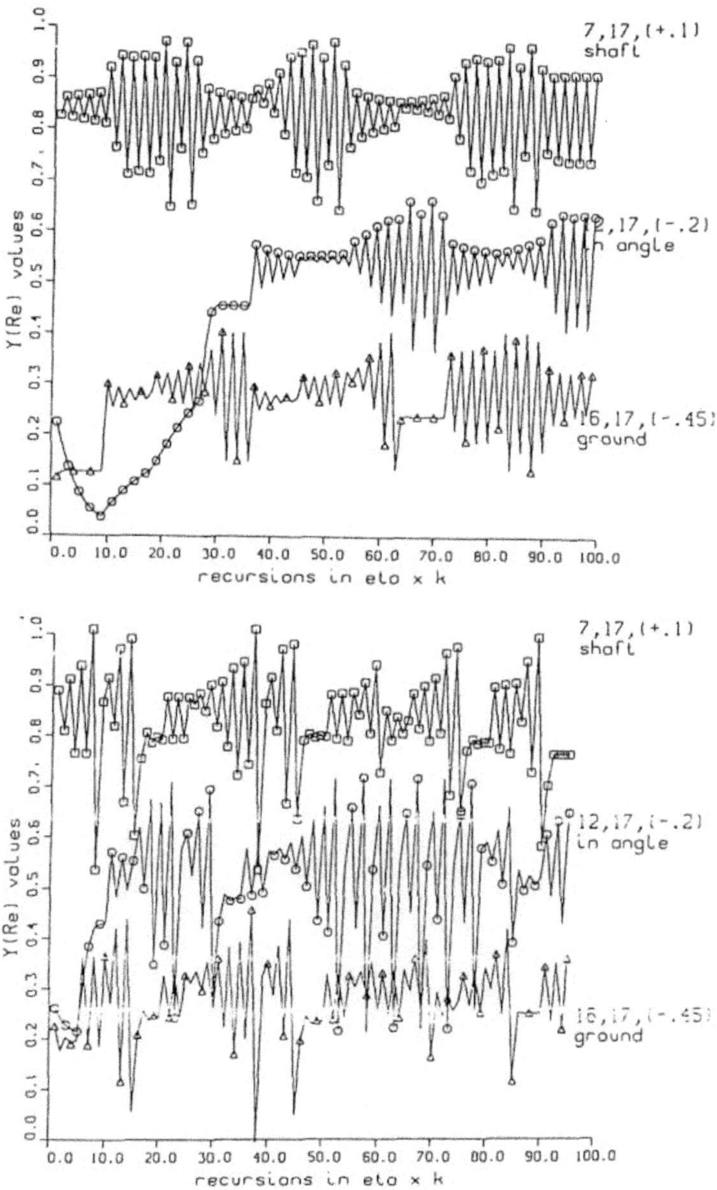

Figures 5.27 and 5.28

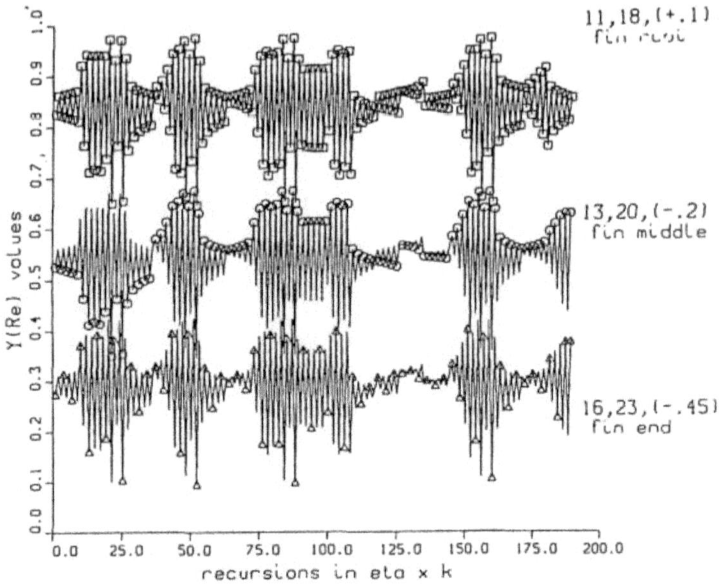

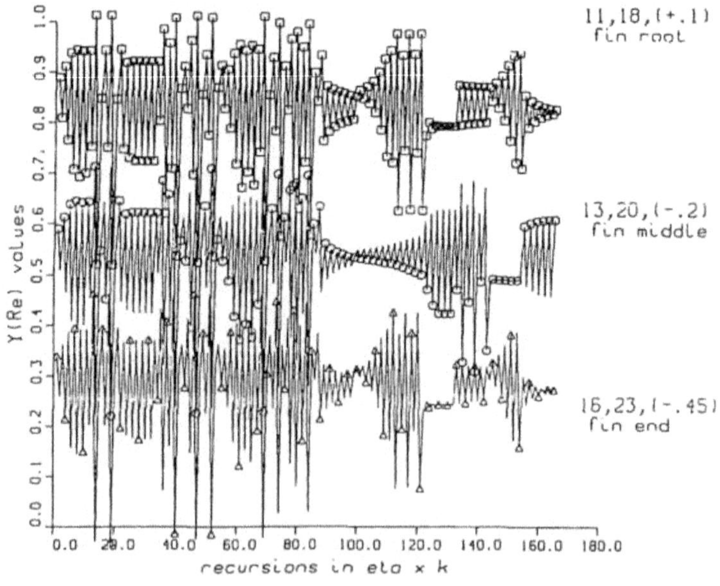

Figures 5.29 and 5.30

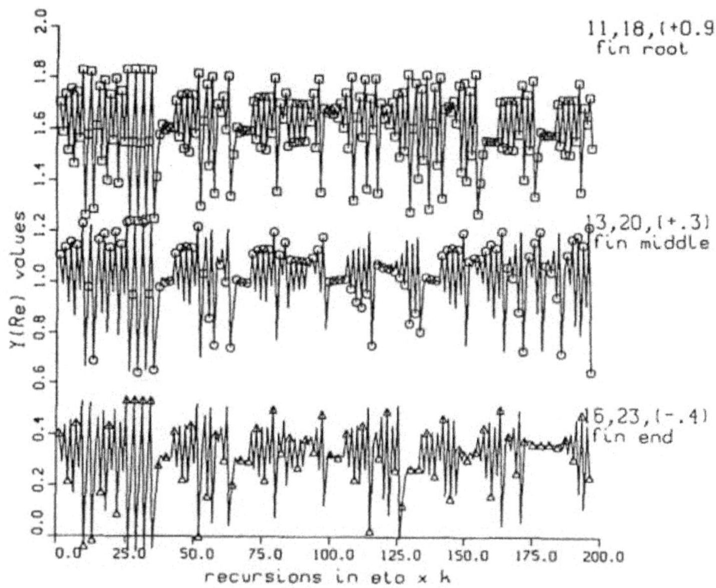

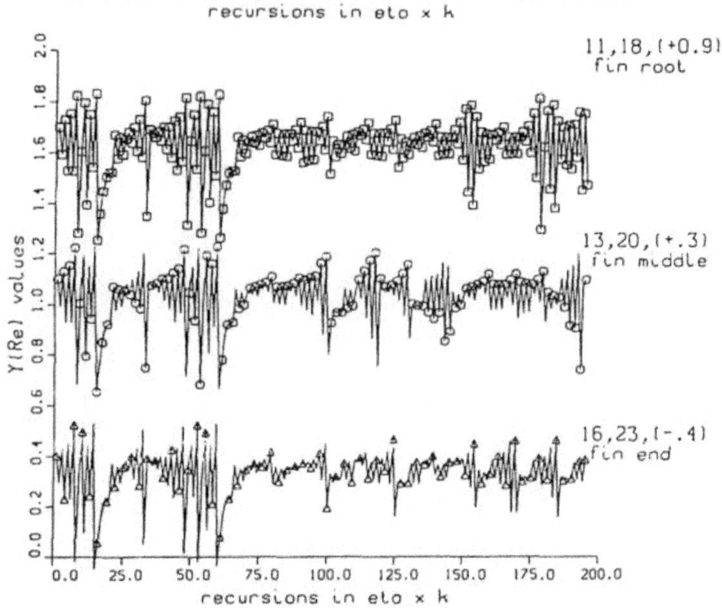

Figures 5.31 and 5.32

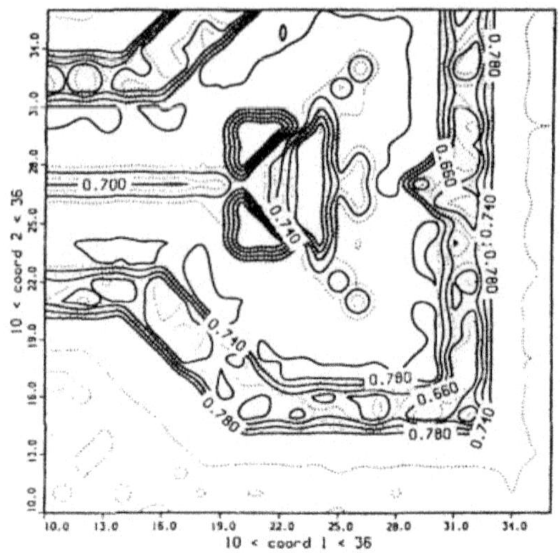
0.700
0.740
0.740
0.780
0.780
0.660
0.780
0.660
0.740
0.780
0.740
0.780
10 < coord 1 < 36
10 < coord 2 < 36

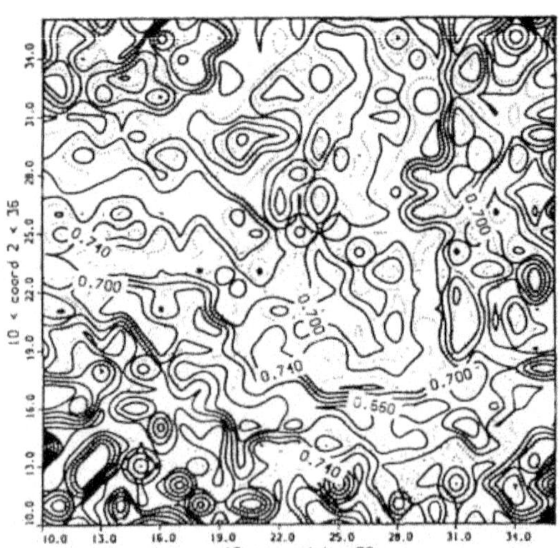
0.740
0.700
0.700
0.740
0.700
0.660
0.740
0.700
10 < coord 1 < 36
10 < coord 2 < 36

order to span, for example, at *the same scale*, a complete display incorporating both the whole Müller-Lyer figures as usually portrayed in a textbook. A neural network with 10^7 or more elements does not necessarily function as a field process with a finer resolution than it needs to serve the organism's purposes in rapid perception, it can clump into a coarser network, functionally but not anatomically, and still induce the effects simulated here.

A curious analogy of the syncronicity properties explored has emerged in the theory, advanced by Crick (1993), that a precondition of consciousness is synchrony of neurophysiological activity over quite wide functional groups; if this speculation is relevant here then perhaps it implies that the $(n \times n)\Gamma^k$ process can be used as a filter which continously processes input, yet admits to consciousness only a subset of pixel activity.

Notes

(1) The points at which a Γ recursion may bifurcate were first examined in Gregson (1988).

(2) Discussions of field theory and induction between figure and ground are used in the context of art by Arnheim (1960,1974) and Gombrich (1960), both of whom are explicitly deriving their analyses from Gestalt psychologists. A purely graphical suggestion for a field and vector theory of illusions and illusory contours is offered by M. Sambin in Petry and Meyer (1987). The term *anomalous contours* is preferred to *illusory contours* by some modern Gestalt theorists, such as Kanizsa (1979). This somewhat metaphysical distinction is not crucial to our present investigations.

(3) In some theories of temporal sensitivity in vision (Watson, 1986), two mechanisms are postulated, one for transients and the other for sustained inputs. Transients are not modelled in this chapter, but it may be noted that the Γ trajectories are sustained in their Real part and transient, for a pulse input, in their Imaginary part. It is possible, but not reported here, to model fields generated by short-lived transients by using $\Gamma V1$ in $(n \times n)\Gamma^k$ and the results are weak flickering figure patterns on a ground that stands out from its surround.

(4) Ott (1993, chapter 8) reviews chaotic transients which resemble the shifts in synchrony pictured here.

6 More Fields generated from Lattices in (n x n)Gamma, and the Psychology of Some Illusions

"..dynamic networks are capable of producing orientation tuning interaction effects and perceptual illusions which have been attributed to more fixed "detector systems".

T. Caelli, (1984, p. 160)

In this chapter we use and extend the results of the previous analysis to look at phenomena for which experimental evidence exists; in many cases the origins of such evidence go back well over a century, and sometimes the detail needed to compare theory with data lies in older studies, because of the 20th century practice of summarising statistical 'significances', particularly in analysis of variance, as a substitute for discursive reporting of results. The use of the general linear model is psychometrics is not necessarily compatible with the mapping of fields generated by nonlinear processes, crucial information is filtered and lost.

Alternative Field Representations

Field theory has been used most systematically in psychology by those working in the tradition of Gestalt theory in order to explain the visual and haptic illusions which are forms of non-veridical perception. The two illusions most explored, with a host of variations in their design, are shown in Figures 6.1a, 6.1b, and Figure 6.1c is a special case devised by Rubin (1950) which raises serious problems for any theory utilising continuous fields and forces. Some arguments advanced to explain the form and extent of illusions have been:

(a) angle distortion,

(b) line orientation distortion,

(c) line segment length distortion,

(d) induced pseudo-perspective,

6.1a

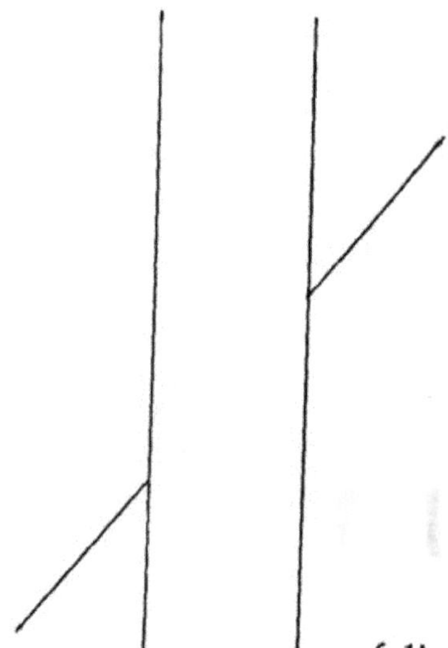

6.1b

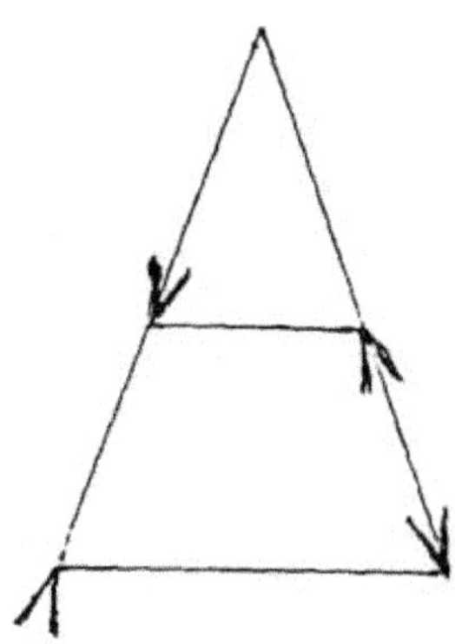

6.1c

Illusion Figures: 6.1a, the Müller-Lyer; 6.1b, the Poggendorf; 6.1c from Rubin (1950)

(e) local ambiguity of the figure's geometry,

(f) relation to a natural vertical-horizontal grid,

(g) subjective contours and line continuation,

(h) field theories, Gestalt and qualitative,

(i) neural networks with mixed inhibitory and excitatory nodes.

Clearly some of these are more descriptions of what happens at output than process explanations, and there is some overlap in the implied mechanisms. Some may be subsumed under the expected consequences of field dynamics, and some are at best *post hoc.* The neural network approach is most recent, because it relies on computer simulation, whereas the previous ideas (a to g) can be, and were, derived from a patient examination of the fine-grain details of psychological data. Both (h) and (i) have been related to possible underlying physiological events in the visual system, and in that sense are reductionist. The most thorough review of previous experimental psychophysical results and theories, from circa 1850 up to 1966, is by Rausch (1966), who also groups the Müller-Lyer (Figure 6.1a) and the Poggendorf (Figure 6.1b) illusions causally together, and treats the eventual percepts as dynamic compromises when he comments (p. 796) "...einer figural-optischen Dynamik und eines Zussamenhangs zwischen Metrik und Auswirkung von Kräften in die Wahrnemungslehre eigeführt hat". At the same time this dynamic process has its limits in what it can produce, (p. 802) "Der Entzerrung sind Grenzen gezogen; sie erfolgt nicht vollständig...". His review necessarily covers Gestalt field theories, and uses the metaphor of a force field in some descriptions of induction between figure and ground.

Robinson (1972) gives the most comprehensive coverage in English of theories about illusions.

It is pertinent to remark that field theories with evolutions governed by integro-differential equations, or networks with generation over adjacent elements, have existed in various forms in vision research for over a century. Ratliff (1965, part 1 chap. 3) reviews six models, built from either linear, nonlinear, or simultaneous linear equations, originating betwen 1865 and 1956. He noted (p. 141) that the common feature of all the models is opposed excitatory and inhibitory influences. The $(n \times n)\Gamma$ networks to be used here are, however, built from simultaneous cross-coupled nonlinear equations (because they are written in complex variables) and have no units in a lattice which can be partitioned into inhibitory and excitatory subclasses[1].

[1] Also, the NPD approach is not written to simulate retinal structures, but

Interest is therefore in parsimony of explanation across diverse psychophysical contexts; extension into visual illusions here is a new area of application. No new model is being created for that purpose. Indeed, it is not plausible that $(n \times n)\Gamma$ is a retinal model, but rather a total system analogue. The use of networks or lattices as the substrates upon which stimulus inputs are mapped has been previously explored; one example is that by Wiles, Lippiatt and Broerse (1990). In that example a 5×5 lattice was set up, but at each pixel there lies a vector of four directionally sensitive elements. This sensitivity was respectively to vertical, horizontal, $+45^o$ and -45° directions[2]. The resultant potential, in magnitude and direction, at each pixel is then dependent on the activation levels induced in the cells of that local vector. Wiles *et al* approached the problem as one of line completion, in analogy with some other visual illusions, and explored the consequences of moving the two vertical parallels of the Poggendorf figure together and apart. This induces some hysteresis in the extent of the illusion. The neural network they employed could be trained to produce an analogue of the Poggendorf illusion. The question of interest here is whether the multiple element complexity is necessary; in $(n \times n)\Gamma$ there are only two processes at one pixel, the parts of the complex variable Y, and these have no directional sensitivity.

There is a conceptual problem with the Wiles *et al* (1990) approach, in that their neural network learns to produce the analogue of the illusion. In psychophysical data the Poggendorf illusion (like many others) is immediately given on initial stimulus presentations, and can dissipate with repeated presentations, which is rather the reverse of the network dynamics through time. This is provided that the figure is actually stable; if parts fluctuate relative to the two parallels, the illusion is attenuated (Mori, 1981). That implies the initial presentation should not be tachistoscopic but have some small duration for the illusion to build up.

A quite different approach, which also implicitly relies on field ideas, was that attacked, experimentally rather than by simulation, by Coren and cowork-

---

rather central higher-level processes. However, the action of λ/a $(= e)$ in $n\Gamma$ does resemble the reciprocal inhibition between hypothetical neurons which, as noted by Ratliff (1965, p. 77) is a ubiquitous feature of the models he reviewed.

[2] The much quoted work of Hubel and Wiesel (Hubel, 1963) on the existence of direction-sensitive neurons in the visual cortex was perhaps a reason for postulating such a structure; later work suggests that in fact direction-sensitivity is supported not by single neurons by by small local networks.

ers (Coren and Girgus, 1972, Ward and Coren, 1976, Coren, Ward, Porac and Fraser, 1978). Here the stimulus figure was blurred and the consequences of the degree of blurring showed to affect the extent of the illusion. Blurring is, of course, a sort of local diffusion, and can be a smearing or an induction of collateral bands of secondary images around the periphery of a figure, as in the case of Mach bands at an edge contour. The idea advanced by Coren is that the visual receptor system is optically imperfect and goes slightly out of sharp focus, inducing blurring; the higher-level cortical networks then are supposed to resharpen this image. Image retrieval by deblurring networks is now well-developed in computer-based transformations (for a review see Vardi and Lee, 1993).

In the present context it is worth noting what would seem to be trivial or irrelevant in most accounts of visual illusions based on plane figures; namely that the illusions are produced when the line of sight is approximately orthogonal to the plane of the figure. If we draw the Poggendorf figure on a sheet, and then view the sheet at an angle where the line of sight is almost parallel to the sheet surface, and we view the figure along the direction of the broken line, then the illusion vanishes. However, if the figure is viewed orthogonally as intended in normal scanning behaviour, it is reported to persist under rotation of the figure in the plane of the sheet. This supports the line continuation mechanism explanation, but does not support, nor yet necessarily contradict, the blurring explanation. However, we may choose not to regard line continuation as a primitive mechanism, and the use of connected lattices is a way of addressing precisely that point. It is probably fruitless to seek any single mechanism to cover the visual illusions, as demonstrations invalidating generalised theories are plentiful and progress in theory construction since the 1860s has been somewhat unrewarding (Over, 1968, Robinson, 1972). Perhaps it is germane to note that Piaget's (1961) centration theory of visual illusions also implies a notion of a field.

The Role of Acute and Obtuse Angles

A strictly limited and specific question, which was addressed in the previous chapter, was; can the role of line segment junctions, with their included angles, as only one mechanism in the generation of illusions, be derived from the lattice form $(n \times n)\Gamma$, though that structure has no defined or constructed *a priori* relation to geometrical illusions nor to the physiology of retinal cell structures?

A common feature of both the Müller-Lyer and the Poggendorf figures

---

Some Qualitative Observations

Some key observations of Judd's are listed here, as a precursor to seeing if any properties of $(n \times n)\Gamma^k$ evolutions, or intermediate stages in such evolutions, appear to echo the phenomenology carefully noted by Judd (1899).

(p. 242) "Overestimation within the figure is accompanied by underestimation of the neighbouring space".

(p. 244) "The point half-way betwen two unequal figures will be attracted towards the larger figure".

"To say that points are shifted in their spatial relations with reference to all the points in the field of vision would be much nearer to the whole truth".

(p. 252) "We cannot appeal to the principle that filled space is overestimated when compared to empty space".

(p. 253) "The more intense the sensation of movement, the greater will be the estimation of the distance. Conversely, the less the intensity of the sensations of movement, the shorter the estimated distance".

(p. 254) "A characteristic observation.....while comparing the horizontal lines with the interval between the free ends of an interrupted horizontal. The ends of the lines and of the interval seemed more or less uncertain".

(p. 257) "When horizontal or vertical distances along the parallel lines are the subjects of attention, as long as they are in the usual positions of the Poggendorf figure, those portions of the parallels lying on the obtuse-angle side of the intercepted line witll be overestimated".

(p. 259) (on the judgment of angles) .."the whole process is a kind of triangulation".

---

is that acute and obtuse angles at line segment junctions play some critical role in inducing the extent of the illusions[3]. Indeed, acute angles seem to play a dominant role in the perception of patterns which generate illusions; they are "powerful attractants of fixation" (Robinson, 1972, p.168). If the angles between line segments are set at 90° or at 0° then the illusions are destroyed.

---

[3] There are varieties of the Müller-Lyer where the ends are bent lines or circles, involving no acute angles, but they do involve line junctions; other variations use dot patterns and no physically-present junctions. These will be discussed as field induction problems later.

There are other second-order complications; for example the thickness of drawn lines in terms of angle subtended at the eye, as compared with the whole figure, may affect the perception of the location of the effective end of a line segment. If the figures are incorporated within meaningful pictures, such as landscape sketches, the illusions may be attentuated. The extent of the illusions is not stable over sustained perception, and can vanish for a fixed display (Judd, 1902). Coren (1970 et seq) showed that altering the shape of the Poggendorf figure at the angle junctions, so that the oblique line made tangent contacts with the vertical lines could destroy the illusion. The various demonstrations that the illusion is induced by the configuration of lines lead to the theory that lateral inhibition of adjacent loci in the visual field was a sufficient cause of the illusions, but this is now known not to be enough. Both retinal and central processes appear to be implicated, so that the illusions are a total system phenomenon, and the presence of feedback in that system makes it rather meaningless to attempt to quantify, as in linear regression, the relative contribution of segments or stages of the optical pathway. Even so, if acute angles are filled in, in the region within the angle, by blurring or by the creation of secondary images, this is enough to produce some distortion which resembles the illusions, provided always that viewing is orthogonal to the plane of the figure.

Let us call the acute angles included in the Poggendorf and Müller-Lyer illusion figures θ; it is assumed that, as is often the case, all acute angles are drawn equal to θ. If $\theta = 0^o$ the figures vanish, because lines become superimposed. If $\theta = 90^o$, the illusions vanish, because the two contrasted figures become identical in the Müller-Lyer case, and become a crossing (like a railway line under a road bridge on a simple map) for the Poggendorf figure. It appears that we could maximise the extent of the illusions by varying θ. Attempts have been numerous (Robinson, 1972, p. 28) to find an optimum value of θ to maximise the illusory effects, and no consensus has been reached on what it might be. In nonlinear systems there is no reason to expect a single optimum value.

The second question for consideration is the orientation of the figures relative to the observer's vertical. Usually the Poggendorf illusion is induced with the two parallel lines vertical, and the Müller-Lyer with the two central parallel lines horizontal. What happens as the figures are rotated through 180° is not clear, but it has been partly investigated (Weintraub and Virsu, 1971), and the perception of acute angles varies with orientation. If the illusions are dependent upon orientation (the receptor matrix is then direction sensitive) then direction should matter, at least to second-order effects. If vertical and horizontal have some special priority in a visual system (Asch and Witkin, 1948) then it may be that the ways in which both Poggendorf and Müller-Lyer figures are usually shown to the observer are fortuitously optimal. Robinson (1972) cites evidence that rotating the Poggendorf illusion so that the oblique line becomes horizontal and the two parallel lines oblique will diminish the illusory effect.

Given that the illusions are produced by a nonlinear system functioning as an integrated whole, the question addressed here is whether the properties of the $(n \times n)\Gamma^k$ model resemble the illusions any better than the other field theories that have been advanced. There are two phenomena that could be generated by the $(n \times n)\Gamma^k$ process, blurring and line completion; outputs have to be examined to see if these occur, and how. First it is necessary to explore the evolution of a pattern, given the variant and its parameters, over a_0, a_f, λ, k, c_1, and c_2.

Simulated Examples

Examples, always with the same core parameter settings of a_o, a_f, η, k, c_1, as illustrations of what can happen in **W1** through **W4**, are first described in Figures 6.2 through 6.10.

In each case the line figure is a vertical line with a branch line segment at 45° inclined upwards. All the figures have had lattice edge effects suppressed, by not plotting the edge regions!

This represents the focal structure in the illusions already discussed. **W4** is the richest structure, so it is interesting first to show that acute angle filling (expressed with a contour plot using DISSPLA) does appear in this variant. The parameters used were $a_0 = 3.4, a_f = 4.2, \lambda = .77, Y_0 = .5, c_1 = 3.0, c_2 = 1.5, k = 9$(cascades). In fact it can be shown that, whilst admittedly there is intricate interplay in the effects of the parameters, the crucial parameter in producing angle filling through cascades is c_2 in **W3** and **W4**. The observed values in the plots are $Y_{k,xy}(Re)_\eta$; the actual Y values differ little over the lattice and it is the relative levels that are important. Actually the local region inside the angle is not sharply different from the induced a_0 values at the location of the imposed line figure $a_f = 5.60$. This is shown in Table 6.1.

Table 6.2 shows an earlier evolution, for $k = 7$, of the same process but with a different figure; lines at 90°. The flip-flop of levels in the lattice from one cascade to another had already been noted in $(n \times n)\Gamma$ analyses of symmetry preservation (Gregson, 1992a). In Table 6.2 the a_f values have become 5.17, which is lower than their immmediate neighbours. The spreading, which

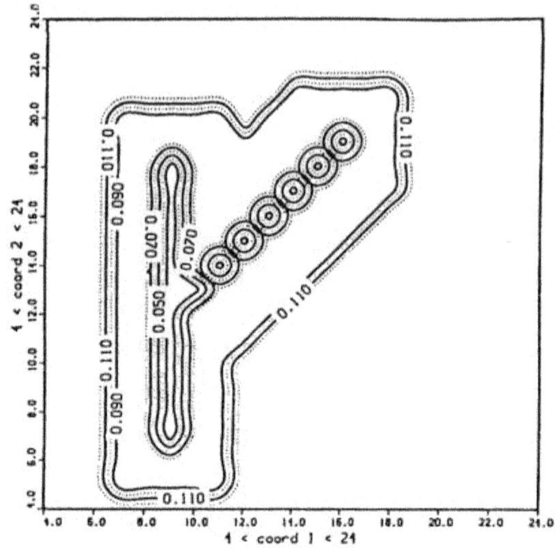

Figure 6.2:
5×5 **W2** $\eta = 5, \lambda = .77, a_0 = 3.4, a_f = 4.2, k = 9$, one diffusion boundary, no angle filling.

corresponds to the Mach band phenomena at least in appearance, with hills and valleys, is shown by the envelope contours[4]. In **W2** with the same core parameters, there is some spread but it terminates abruptly, and there is no filling of the acute angle.

Three other series of fields have been created; they serve to illustrate different effects induced by the proximity of linear pixel arrays serving as line segments. All the examples chosen have precedents in the construction of figures used in perception experiments, in the work of, for example, Judd (1899) and others reviewed by Robinson (1972). They also raise nice questions concerning the generation of Gestalt phenomena such as the grouping of dots into clusters, and the suppression or enhancement of local features. These exam-

[4] It should be noted that there is no Fourier analysis involved in this study.

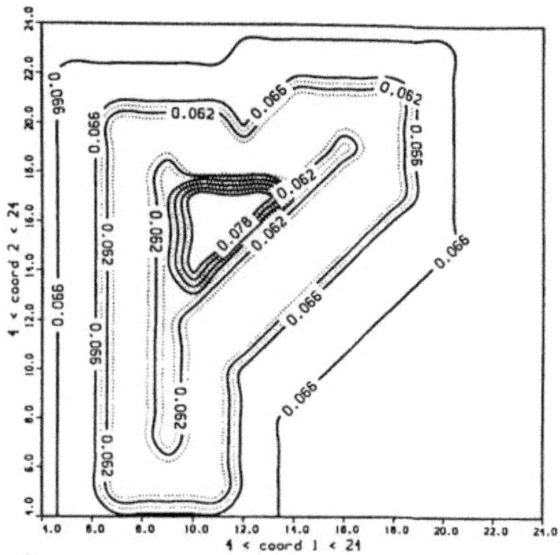

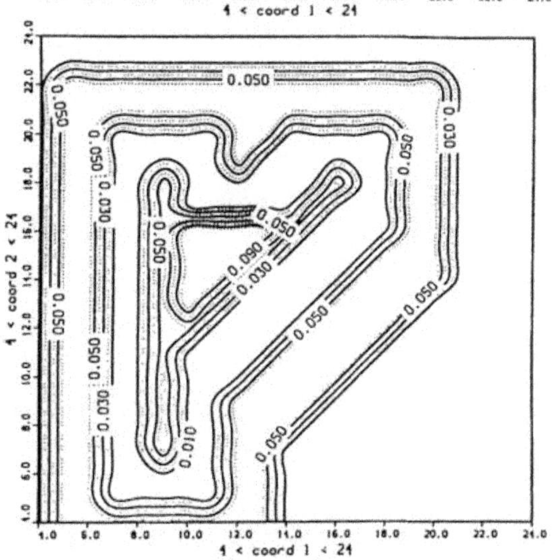

Figure 6.3:

5×5 **W4** $\eta = 5, \lambda = .77, a_0 = 3.4, a_f = 4.2, k = 9$, two bounds and angle filling; contrast with previous figure, and:

Figure 6.4:

5×5 **W4** $\eta = 5, \lambda = 1.1, a_0 = 2.8, a_f = 4.8, k = 9$, more bounds with higher λ and more figure-ground contrast, otherwise the same as Figure 6.3.

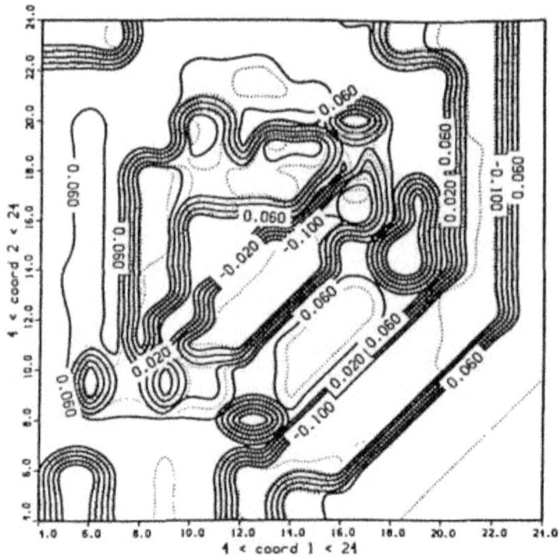

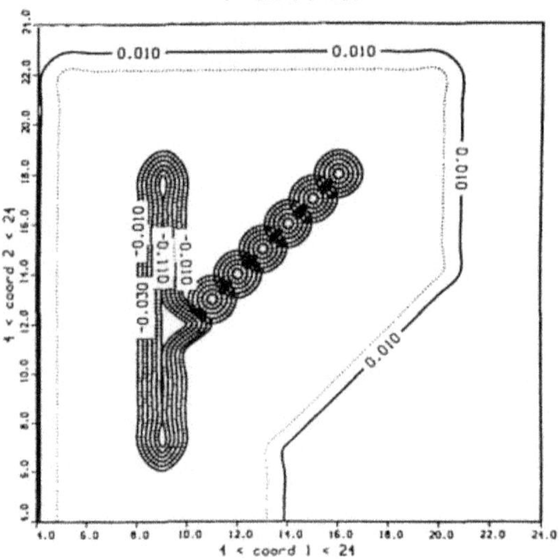

Figure 6.5:

5×5 **W4** $\eta = 5, \lambda = 1.1, a_0 = 2.8, a_f = 4.8, k = 19$, longer cascading of the same process as the previous figure, and:

Figure 6.6:

5×5 **W4** $\eta = 5, \lambda = 1.1, a_0 = 4.8, a_f = 2.8, k = 9$, the negative image case, a_0 and a_f reversed; no angle filling appears at $k = 9$.

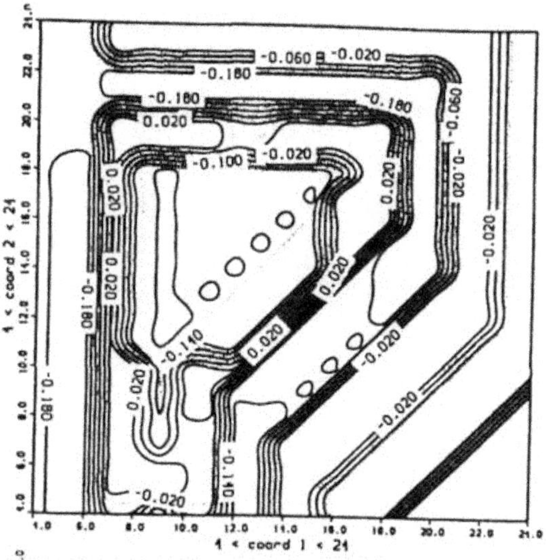

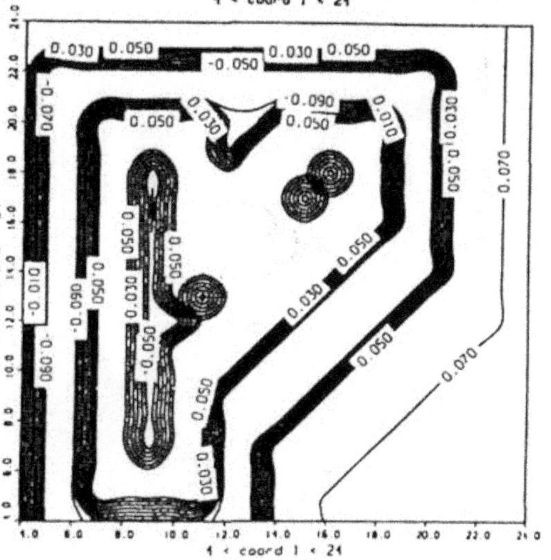

Figure 6.7:

5×5 **W4** $\eta = 5, \lambda = 1.1, a_0 = 4.8, a_f = 2.8, k = 19$, the negative image case, a_0 and a_f reversed; later cascade, compare with Figure 6.5, and:

Figure 6.8:

5×5 **W4** $\eta = 5, \lambda = 1.1, a_0 = 2.8, a_f = 4.8, k = 9$, positive image but with a spot at $a = 3.8$ introduced in the angle's concave region. The suppression of part of the figure in the angle's 45° branch is marked.

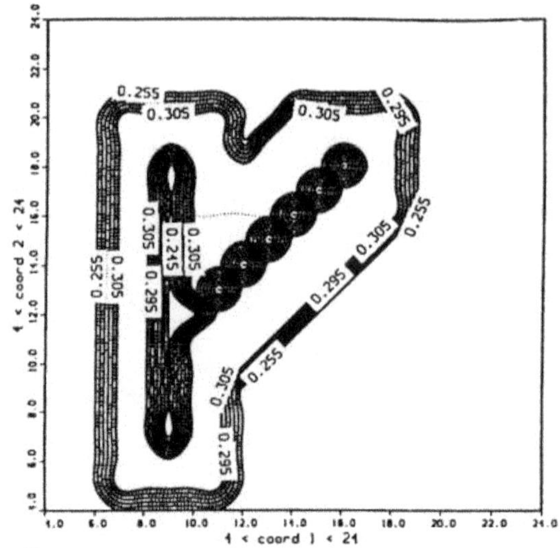

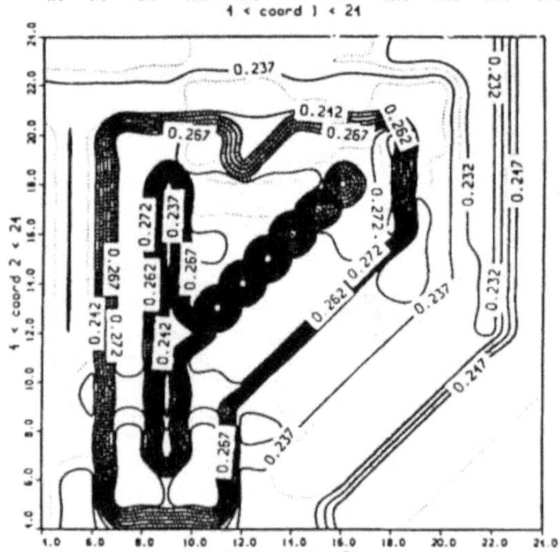

Figure 6.9:

5×5 **W4** $\eta = 5, \lambda = 1.1, a_0 = 2.8, a_f = 4.8, k = 9$, entropy figure over the first 9 cascades, ghost of angle filling appears, and:

Figure 6.10:

5×5 **W4** $\eta = 5, \lambda = 1.1, a_0 = 2.8, a_f = 4.8, k = 19$, entropy figure over the first 19 cascades, turbulence over the field appears, input figure is still dominant.

Table 6.1

Central part of the matrix of a values for $k = 9$ with the 45° branch figure

5.14	5.14	5.14	5.14	5.35	5.35	5.30	5.30	5.30	5.30	5.30	5.30
5.14	5.60	5.14	5.14	5.35	5.35	5.30	5.30	5.30	5.30	5.30	5.30
5.14	5.60	5.14	5.14	5.35	5.35	5.35	5.30	5.30	5.30	5.30	5.30
5.14	5.60	5.14	5.14	5.35	5.35	5.35	5.35	5.30	5.30	5.30	5.30
5.14	5.60	5.14	5.14	5.14	5.35	5.35	5.35	5.35	5.30	5.30	5.30
5.14	5.60	5.14	5.14	5.14	5.14	5.35	5.35	5.35	5.35	5.30	5.30
5.14	5.60	5.60	5.14	5.14	5.14	5.14	5.35	5.35	5.35	5.35	5.30
5.14	5.60	5.14	5.60	5.14	5.14	5.14	5.14	5.35	5.35	5.35	5.35
5.14	5.60	5.14	5.14	5.60	5.14	5.14	5.14	5.14	5.35	5.35	5.35
5.14	5.60	5.14	5.14	5.14	5.60	5.14	5.14	5.14	5.14	5.35	5.35
5.14	5.60	5.14	5.14	5.14	5.14	5.60	5.14	5.14	5.14	5.14	5.35
5.14	5.60	5.14	5.14	5.14	5.14	5.14	5.60	5.14	5.14	5.14	5.35
5.14	5.60	5.14	5.14	5.14	5.14	5.14	5.14	5.60	5.14	5.14	5.35
5.14	5.14	5.14	5.14	5.35	5.14	5.14	5.14	5.14	5.14	5.14	5.35
5.14	5.14	5.14	5.14	5.35	5.35	5.14	5.14	5.14	5.14	5.14	5.35
5.35	5.35	5.35	5.35	5.35	5.35	5.35	5.35	5.35	5.35	5.35	5.35

Table 6.2

Central part of the matrix of a values for $k = 7$ with the 45° branch figure
Same parameters as in Table 6.1, but earlier in the process.

5.32	5.32	5.32	5.24	5.21	5.21	5.21	5.21	5.21	5.21	5.21	5.21
5.32	5.17	5.32	5.24	5.21	5.21	5.21	5.21	5.21	5.21	5.21	5.21
5.32	5.17	5.32	5.24	5.21	5.21	5.21	5.21	5.21	5.21	5.21	5.21
5.32	5.17	5.32	5.24	5.21	5.21	5.21	5.21	5.21	5.21	5.21	5.21
5.32	5.17	5.32	5.24	5.24	5.21	5.21	5.21	5.21	5.21	5.21	5.21
5.32	5.17	5.32	5.32	5.24	5.24	5.24	5.24	5.24	5.24	5.24	5.21
5.32	5.17	5.32	5.32	5.32	5.32	5.32	5.32	5.32	5.32	5.24	5.21
5.32	5.17	5.17	5.17	5.17	5.17	5.17	5.17	5.17	5.32	5.24	5.21
5.32	5.17	5.32	5.32	5.32	5.32	5.32	5.32	5.32	5.32	5.24	5.21
5.32	5.17	5.32	5.24	5.24	5.24	5.24	5.24	5.24	5.24	5.24	5.21
5.32	5.17	5.32	5.24	5.21	5.21	5.21	5.21	5.21	5.21	5.21	5.21
5.32	5.17	5.32	5.24	5.21	5.21	5.21	5.21	5.21	5.21	5.21	5.21
5.32	5.17	5.32	5.24	5.21	5.21	5.21	5.21	5.21	5.21	5.21	5.21
5.32	5.32	5.32	5.24	5.21	5.21	5.21	5.21	5.21	5.21	5.21	5.21
5.24	5.24	5.24	5.24	5.21	5.21	5.21	5.21	5.21	5.21	5.21	5.21
5.21	5.21	5.21	5.21	5.21	5.21	5.21	5.21	5.21	5.21	5.21	5.21

ples are by no means exhaustive, and with any series intermediate cases can be constructed by a choice of $\eta, k, \lambda, a_0, a_f$. Parameter settings used are given for most examples; contour maps are mainly of $Y_{xy,k}(\mathrm{Re})$ values and not of B as noted.

The figures 6.11 through 6.31 illustrate some characteristic properties induced in the contour maps by particular figure-ground input configurations, over narrow parameter windows. The broken parallel line segments (figures 6.11 - 6.16) were chosen because they facilitate the generation, under restricted parameter settings, of line continuation in the ground between collinear segments. They can also induce contours orthogonal to the direction of the segments. This has physiological parallels (Heydt, 1987, p. 131) in the action of cells in cortical areas V1 and V2; it is worth noting here that there is evidence against Fourier transforms of filtered images constituting a basis for orthogonal illusory contours[5]. Contour completion and contour induction at the boundaries and interfaces of incomplete patterns are a ubiquitous features of perception, though such contours are often phenomenologically weak and unstable as compared with the figures themselves, and they have been invoked as a mechanism to generate some illusions. The creation of new line segments, echoing pixel strings lying on axes of symmetry and adjacent to the input, can sometimes arise, but in phenomenological accounts this would appear to be a much rarer event.

Figures 6.17 - 6.22 are based on a concave line figure ([) made from three line segments. This serves as an alternative to the arrow forms of chapter 5; much the same dynamics arise, the only difference is that angles are now right and not acute or obtuse, but the filling still diffuses rapidly into local concave regions and induces effects on both sides of a ⊣ junction of line segments.

The figures 6.23 - 6.29 are in some ways more interesting, because here a set of 10 dots, arranged in two parallel arrays of five each, is used. But now there are five graded values of a_f; the dots increase in a_f (and hence in figure-ground contrast) as one goes down the figures. This has a number of subtle consequences; some weaker points can vanish, and other stronger inputs form local groupings, which some might like to call *coalitions*. The contours take on intricate patterns, and the two arrays are partitioned into two bounded sets.

[5] van der Heydt (1987, p. 139) comments "I have often taken up a position against the cortical Fourier analysis theory, partly because of the evidence for nonlinear processes at early stages, but also because nobody seems to know how the cortex might proceed after the extraction of the Fourier components."

This does recall the familiar Gestalt principle of grouping. The point is that, over the whole series of figures used, from one algebraic network equation a diversity of field patterns are generated, but they are not separate independent phenomena even though they can each be given descriptive names[6].

To expand on this latter comment, we revert to the distinction between process and outcome models which was drawn in chapter 3. The form of Gestalt theory which consists of a list of principles, as exemplified in the notes at the end of this chapter giving some of Helson's (1933) summary in English of the Gestalt position[7], is usual. Only those of the 114 points which Helson listed, which appear to be interpretable as statements about figure-ground transformations, are given. Some of his statements are so abstract that they are metatheory which has to be reinterpreted to yield any substantive predictions. The distinction drawn between verbalised Gestalt theory and $(n \times n)\Gamma^k$ lattice properties may be summarised in two alternative global mappings:

Γ : algebraic network model $\Rightarrow$ dynamic process $\Rightarrow$
theoretical output patterns

versus

Gestalt: observed outcomes $\Rightarrow$ catalogue of properties $\Rightarrow$
enumeration of principles

The first term of the second string is postulated to be a subset of the last term of the first string. If there exist observed outcomes which are not predicted, then the Γ theory is incomplete, if there exist predictions for which no data have yet been observed, the theory is potentially falsifiable. Both these mappings should ideally be closed by a return loop through experimental data; the left-hand terms in both are subject to recursive revision. In short, the Γ lattices are generators for an inductive process description, and from the process contour maps of various sorts may be obtained by tapping into the process at points in its evolution, or by smoothing over local cascade sequences. In contradistinction, the Gestalt approach works backwards from observed phenomena to deduce a set of properties which collectively (with or without intrinsic redundancy) constitute a sort of taxonomy of implicit central processes. Every time a new outcome is discovered, the taxonomy is augmented. Consequently,

---

[6] It is even the case that some patterns evocatively resemble the Maori art on the walls of Te Whare Marae, which New Zealanders besides the author will notice.

[7] It is accepted for the present purposes that Helson did give a fair, if terse, summary of the original sources in German.

metatheoretical principles have to be written in an accomodating manner, that makes them vague. The process approach is much tighter in its initial assumptions, yet many potential outcomes can be created, but may never be observed because their parameter settings do not correspond to any biologically realisable settings. The identification of of the actual psychophysics is thus partly a problem of finding the limits of a parameter space within which the dynamics correspond to reality.

Table 6.5

Central part of the matrix of a values for $k = 17$ with the [figure

Using $\lambda = 1.3535$ and $\eta = 7$

4.54	4.54	4.59	4.59	4.59	4.59	4.59	4.59	4.59	4.59	4.58	4.58
4.47	4.47	4.47	4.47	4.47	4.47	4.47	4.47	4.47	4.47	4.47	4.47
4.47	4.47	4.47	4.47	4.47	4.47	4.47	4.47	4.47	4.47	4.47	4.47
5.01	5.01	5.01	5.01	5.24	5.24	5.24	5.24	5.24	5.24	4.47	4.47
5.01	5.01	5.01	5.01	5.24	5.24	5.24	5.24	5.24	5.24	4.47	4.47
5.40	5.40	3.89	3.89	3.87	3.87	3.87	3.87	5.24	5.24	4.47	4.47
5.40	5.40	3.89	4.64	4.86	4.86	4.86	4.86	4.86	4.86	4.50	4.50
5.40	5.40	3.89	4.64	4.91	4.91	4.86	4.86	4.86	4.86	4.50	4.50
5.40	5.40	3.89	4.64	4.91	3.96	3.96	3.96	3.96	3.96	4.50	4.50
5.40	5.40	3.89	4.64	4.91	4.91	4.86	4.86	4.86	4.86	4.50	4.50
5.40	5.40	3.89	4.64	4.86	4.86	4.86	4.86	4.86	4.86	4.50	4.50
5.40	5.40	3.89	3.89	3.87	3.87	3.87	3.87	5.24	5.24	4.47	4.47
5.01	5.01	5.01	5.01	5.24	5.24	5.24	5.24	5.24	5.24	4.47	4.47
5.01	5.01	5.01	5.01	5.24	5.24	5.24	5.24	5.24	5.24	4.47	4.47
4.47	4.47	4.47	4.47	4.47	4.47	4.47	4.47	4.47	4.47	4.47	4.47

Near Chaotic Dynamics and Filling

Because it is already known, from the mathematics of Γ recursions, that quasi-cyclic dynamics leading into chaos in the phase space can be induced with high ae combinations, and it is also known that some brain dynamics are operative at the edge of chaos, it is illuminating to explore how $(n \times n)\Gamma^k$ behaves when parameter values are picked to explore the region in question.

To induce instability within the range which does not lead to computational impossibilities, the contrast $a_f - a_0$ is enhanced, and the λ value pushed up, with exploration to the fourth decimal place, to locate a region near the dynamical bifurcation series. There are, in a cascaded series, two potential

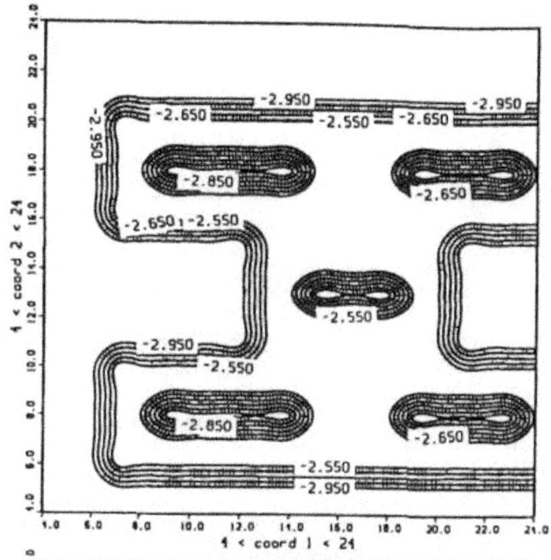

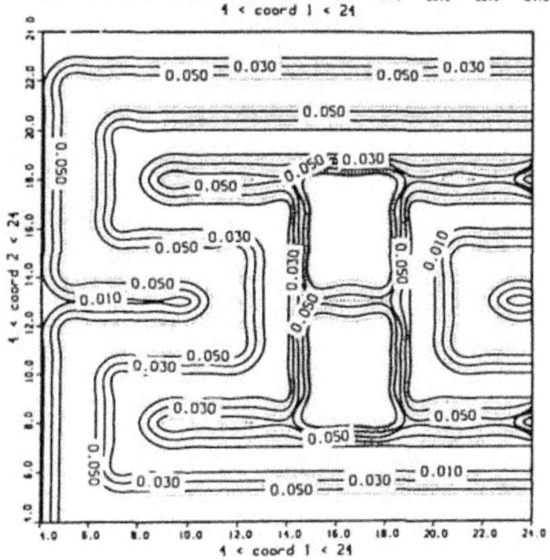

Figure 6.11:

5 × 5 **W4** $\eta = 5, \lambda = 1.1, a_0 = 2.8, a_f = 4.8, k = 11$, cross-entropy figure, original short line arrangement shown. One envelope bound only, and:

Figure 6.12:

5 × 5 **W4** $\eta = 5, \lambda = 1.1, a_0 = 2.8, a_f = 4.8, k = 9$, induction of line segment joining and orthogonal contours in the $Y(\text{Re})$ field.

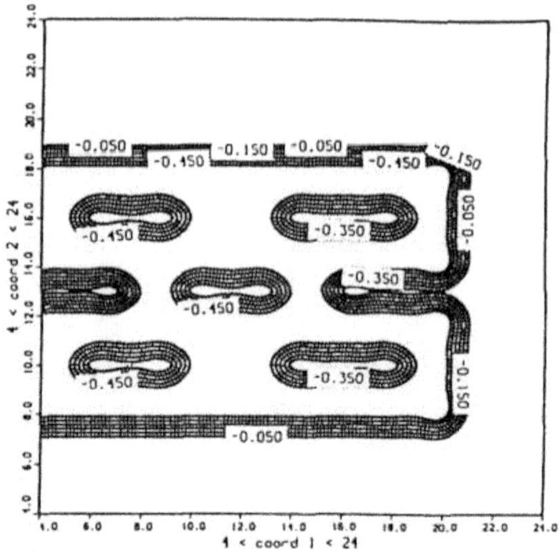

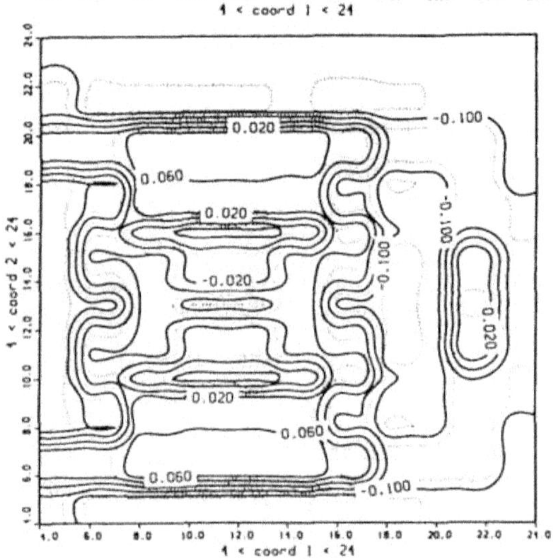

Figure 6.13:

5 × 5 **W4** $\eta = 7, \lambda = 0.8, a_0 = 2.8, a_f = 4.8, k = 9$, lower cross-coupling, longer within-Γ trajectory, and early cascade only induces blurring and 'echoing' in the envelope.

Figure 6.14:

5 × 5 **W4** $\eta = 5, \lambda = 1.1, a_0 = 2.8, a_f = 4.8, k = 17$, approaching turbulent dynamics; original line segments obscured. Subjective contours in collinear gaps dominate; complicated echoing in the periphery.

Table 6.3

Central part of the matrix of a values for $k = 17$ with the [figure
Using $\lambda = 1.3535$ and $\eta = 5$

4.89	4.89	4.89	4.89	4.89	4.89	4.89	4.89	4.89	4.89	4.89	4.89
4.76	4.76	4.76	4.76	4.76	4.76	4.76	4.76	4.76	4.76	4.76	4.76
4.76	4.76	4.76	4.76	4.76	4.76	4.76	4.76	4.76	4.76	4.76	4.76
4.00	4.00	4.00	4.00	4.00	4.00	4.00	4.00	4.00	4.00	4.76	4.76
4.00	4.00	4.00	4.00	4.00	4.00	4.00	4.00	4.00	4.00	4.76	4.76
4.00	4.00	4.48	4.48	4.48	4.48	4.48	4.48	4.00	4.00	4.76	4.76
4.00	4.00	4.48	5.15	5.15	5.15	4.04	4.04	4.02	4.02	4.69	4.69
4.00	4.00	4.48	5.15	5.15	5.15	4.04	4.04	4.02	4.02	4.69	4.69
4.00	4.00	4.48	5.15	5.15	4.87	4.06	4.06	5.14	5.14	4.66	4.66
4.00	4.00	4.48	5.15	5.15	5.15	4.04	4.04	4.02	4.02	4.69	4.69
4.00	4.00	4.48	5.15	5.15	5.15	4.04	4.04	4.02	4.02	4.69	4.69
4.00	4.00	4.48	4.48	4.48	4.48	4.48	4.48	4.00	4.00	4.76	4.76
4.00	4.00	4.00	4.00	4.00	4.00	4.00	4.00	4.00	4.00	4.76	4.76
4.00	4.00	4.00	4.00	4.00	4.00	4.00	4.00	4.00	4.00	4.76	4.76
4.76	4.76	4.76	4.76	4.76	4.76	4.76	4.76	4.76	4.76	4.76	4.76

Table 6.4

Central part of the matrix of a values for $k = 17$ with the [figure
Using $\lambda = 1.3535$ and $\eta = 6$

4.79	4.79	4.79	4.79	4.79	4.79	4.79	4.79	4.79	4.79	4.79	4.79
4.90	4.90	4.90	4.90	4.89	4.89	4.89	4.89	4.89	4.89	4.89	4.89
4.90	4.90	4.90	4.90	4.89	4.89	4.89	4.89	4.89	4.89	4.89	4.89
4.62	4.62	4.64	4.64	4.63	4.63	4.64	4.64	4.62	4.62	4.90	4.90
4.62	4.62	4.66	4.66	4.66	4.66	4.68	4.68	4.64	4.64	4.87	4.87
4.62	4.62	4.87	4.87	4.87	4.87	4.92	4.92	4.64	4.64	4.87	4.87
4.62	4.62	4.87	5.13	4.71	4.71	4.74	4.74	4.97	4.97	4.88	4.88
4.62	4.62	4.87	5.13	4.69	4.69	4.74	4.74	4.97	4.97	4.88	4.88
4.62	4.62	4.87	5.13	4.69	4.83	4.58	4.58	4.72	4.72	4.88	4.88
4.62	4.62	4.87	5.13	4.69	4.69	4.74	4.74	4.97	4.97	4.88	4.88
4.62	4.62	4.87	5.13	4.71	4.71	4.74	4.74	4.97	4.97	4.88	4.88
4.62	4.62	4.87	4.87	4.87	4.87	4.92	4.92	4.64	4.64	4.87	4.87
4.62	4.62	4.66	4.66	4.66	4.66	4.68	4.68	4.64	4.64	4.87	4.87
4.62	4.62	4.64	4.64	4.63	4.63	4.64	4.64	4.62	4.62	4.90	4.90
4.90	4.90	4.90	4.90	4.89	4.89	4.89	4.89	4.89	4.89	4.89	4.89

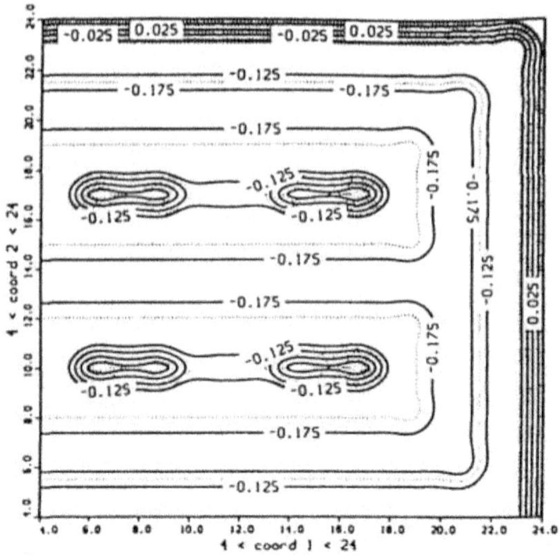

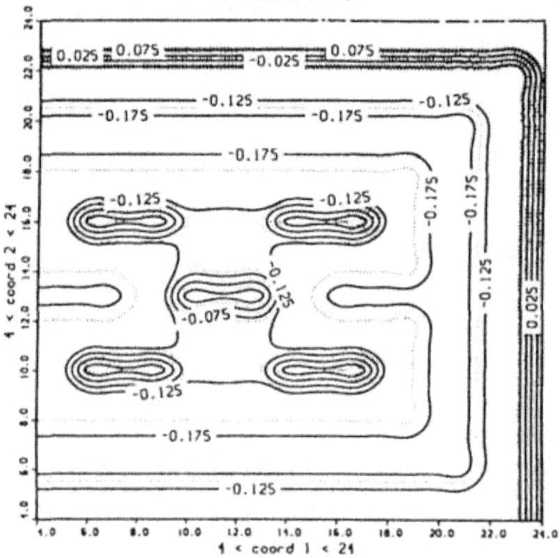

Figure 6.15:

5×5 **W4** $\eta = 7, \lambda = 1.3535, a_0 = 2.8, a_f = 4.8, k = 9$, in the near-chaotic region, induced by the λ value chosen. A simpler line figure, with only four segments; subjective contours and separation bounds induced.

Figure 6.16:

5×5 **W4** $\eta = 7, \lambda = 1.3535, a_0 = 2.8, a_f = 4.8, k = 9$, again in the near-chaotic region, induced by the same λ value chosen. Line figure, with five segments; effect of middle segment shown.

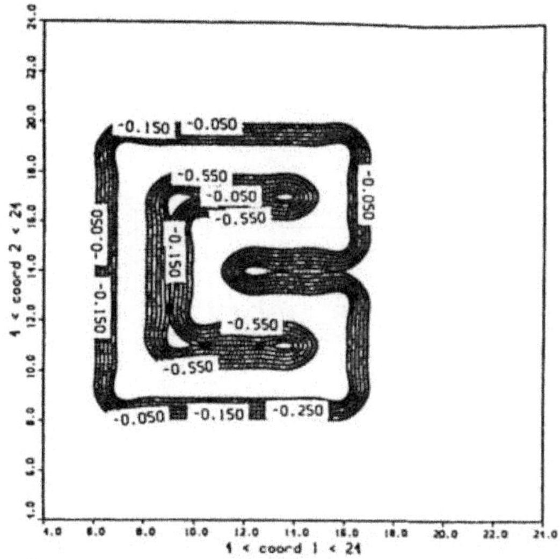

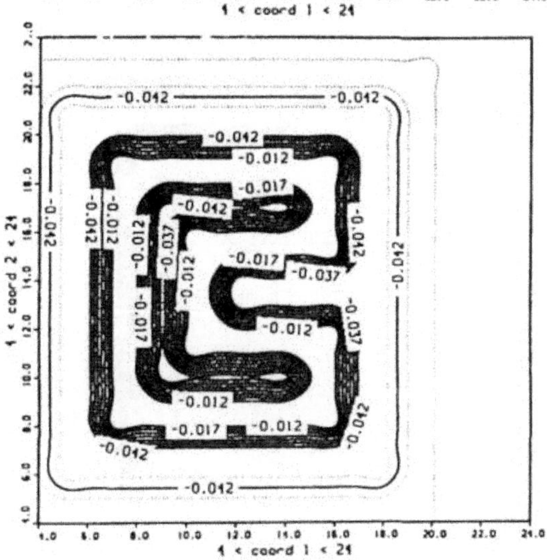

Figure 6.17:

5×5 **W4** $\eta = 7, \lambda = 0.8, a_0 = 3.4, a_f = 4.2, k = 3$, very early stage in the process with one diffusion bound, and:

Figure 6.18:

5×5 **W4** $\eta = 7, \lambda = 0.8, a_0 = 3.4, a_f = 4.2, k = 11$, as before but the diffusion bounds multiply and spread out.

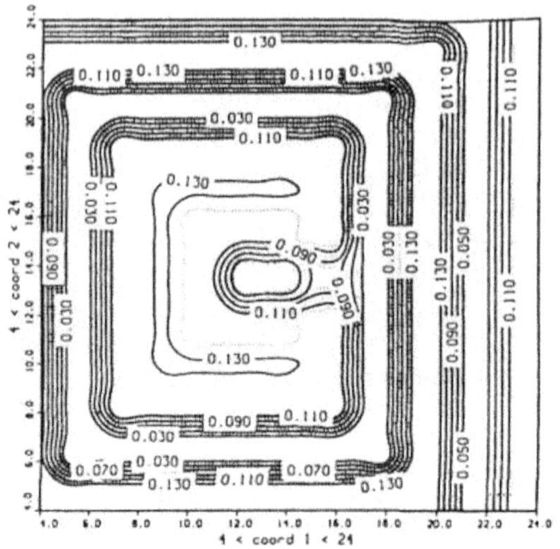

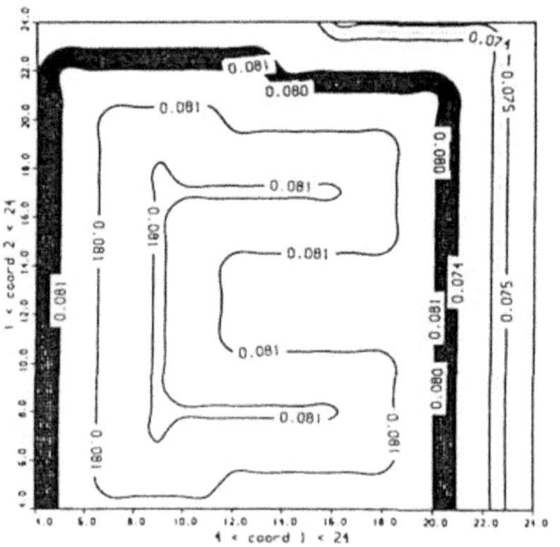

Figure 6.19:

5×5 **W4** $\eta = 7, \lambda = 1.1, a_0 = 3.4, a_f = 4.2, k = 11$, as before, but with λ increased; the appearance of a region on the horizontal symmetry axis within the concave part of the figure, and:

Figure 6.20:

5×5 **W4** $\eta = 11, \lambda = 0.75, a_0 = 3.4, a_f = 4.2, k = 9$, the effect of a longer η and a lower λ, a simpler figure results.

sources of quasi-cyclical outputs, namely fluctatuations with a single Γ trajectory (here corresponding to one cascade step) and fluctuations over cascades k, for fixed other parameters. To illustrate the intrinsic properties of the lattice's dynamics in this parameter region, the [figure which is sometimes used as a line terminator in the Müller-Lyer illusion features in Tables 6.3 to 6.5. A two-dimensional exploration is called for, over a range of λ values for fixed k, and over a range of η values for fixed λ and k. To save space, again only the centre of the matrix is shown, and a values tabled. As the mapping $a \mapsto Y$ is not linear, the corresponding $Y(\mathrm{Re})$ maps are not quite the same, but the object here is to show local instabilities of relatively high frequency in the time evolution (over ηk replications) of the total field process. The unstable dynamics are easiest to illustrate by taking a fairly high value of $k = 17$, though they do emerge for small k. In all the lattices of a shown in Tables 6.3 through 6.5, the initial values were $a_0 = 2.8, a_f = 4.8$, and $c_2 = 1.5$ as before in variant **W4**.

It can be seen that the pattern and degree of filling within the figure is unstable but sometimes marked. The symmetry of the line figure around a horizontal axis is an important property; symmetry is preserved, as found previously (Gregson, 1992a) by $(n \times n)\Gamma$ cascade evolution. This symmetry also induces axis fluctuations which would operate directly on the line of the Müller-Lyer figure if it were present on the convex side of the [figure, but if the line were within the concave side of the figure it would induce more secondary lines parallel to the main real line.

Induced Phenomena

To investigate the dynamics of $(n \times n)\Gamma^k$ more thoroughly, it is expedient to plot outputs, at each cascade k of the matrices of a, $Y(\mathrm{Re})$ and the entropy H and cross-entropy B of the field, as defined in the previous chapter. The entropy is calculated, pixel by pixel, by taking the series of k Y_{xy} values, renormalising each to lie in the interval $0, 1$ over the lattice for that cascade, and then obtaining the mean entropy at that point. The cross-entropy (Boltzmann entropy) is similarly derived, and the normalised initial a matrix is used to give the theoretical values at each pixel.

The evolution over ηk steps in time is the process model, the cross-entropy may be regarded as nearer to an outcome prediction, as it preserves the invariances in the system which are associated with the inputs; in phenomenological terms, it highlights the dominant features of the stimulus display so that they are preserved into the response. If the eventual cross-entropies of the pixels

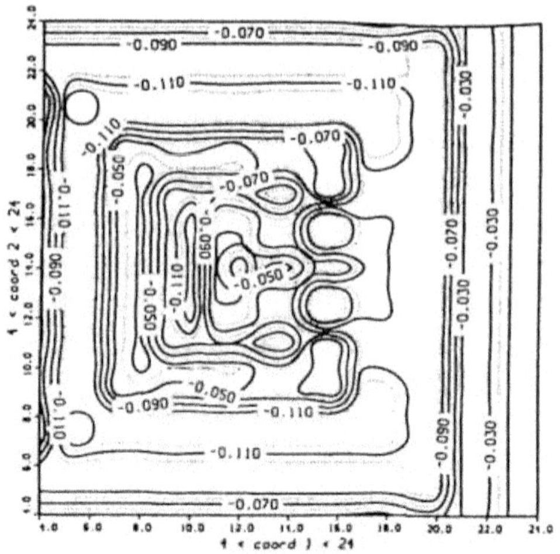

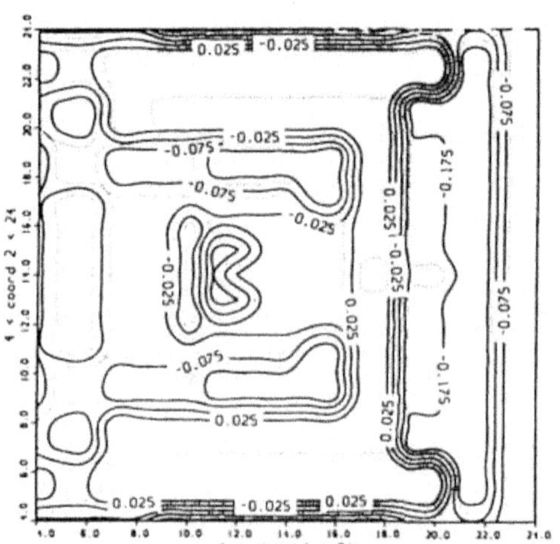

Figure 6.21:

5×5 **W4** $\eta = 6, \lambda = 1.3535, a_0 = 2.8, a_f = 4.8, k = 17$, in the near-chaotic region induced by the λ value chosen. Turbulence particularly within the concave region, but strong symmetry preserved; this implies image uncertainly in the turbulent area, and:

Figure 6.22:

5×5 **W4** $\eta = 7, \lambda = 1.3535, a_0 = 2.8, a_f = 4.8, k = 17$. Because the dynamics have passed the first bifurcation, a very small change in η is sufficient to alter the pattern. The process will flip-flop within a cascade between Figures 6.21 and 6.22.

at the stimulus input locations are minimal compared with the surrounding field, then the image is preserved in the final output. At the same time, the secondary features, such as minimal-convex envelope bounds, angle filling, line continuation, and local inhibition of weaker signals, lie within the process itself.

It is worth remarking that the reported appearance of geometrical illusions is a sharp image displaced, or locally distorted slightly, with perhaps some secondary features at much lower intensity and with associated uncertainty about exactly where parts of the figure are; the process within ηk shows how the secondary features arise and what they may do to the whole image, but these secondary features themselves may not emerge in the final entropy maps. Summarising under headings the main results so far:

Masking: It appears that introducing a single point at a lower a level than the a_f for the line figure's pixels may, with suitable choice of η, λ, induce suppression of some adjacent line segments.

Using a series of single points, with a range of a levels which lie intermediate between a_0 and a_f, produces a diversity of phenomena; weak points may vanish, and convex bounds are created as envelopes around subsets of the stronger inputs. This can partition a spaced bivariate array, and produce the *goodform* clusterings of Gestalt theory.

Orthogonal Local Contours: In some figures the ends of a series of horizontal lines which are in vertical alignment induce a subjective vertical contour which is through those ends. Whether or not this occurs depends on the spacings of the array of input lines.

Line continuation: A break in a line can be filled; A $Y(\mathrm{Re})$ matrix in which the gaps between line segments ($a_{f,k} = 4.48$) is bridged by an induced collinear continuation ($a_{s,k} = 5.18$) in a field of $a_{0,k} = 5.26$ and a local surround to the figure of $a_{s,k} = 5.09$.

Minimum Stimulation and Intermediate Operations

The starting point of inputs here is always a lattice of points; the nearest thing to a line is simply a linear array of a_f pixels. There have been some demonstrations (Day, 1994) that purport to show the Müller-Lyer and Poggendorf figures still induce their illusions if the figure is reduced to a skeletal pattern of dots[8] as shown in Table 6.6.

[8] The assumption made by some theorists is that dot patterns remain functionally invariant to some earlier stages in processing, whereas line segments might not. See comments by Zucker, in Dodwell and Caelli, (1984, p 285).

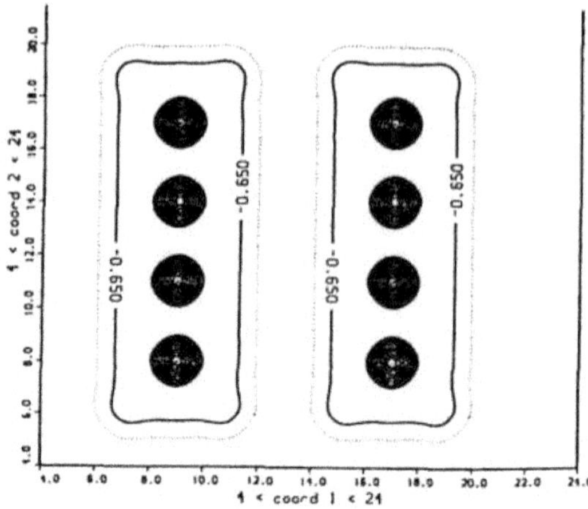

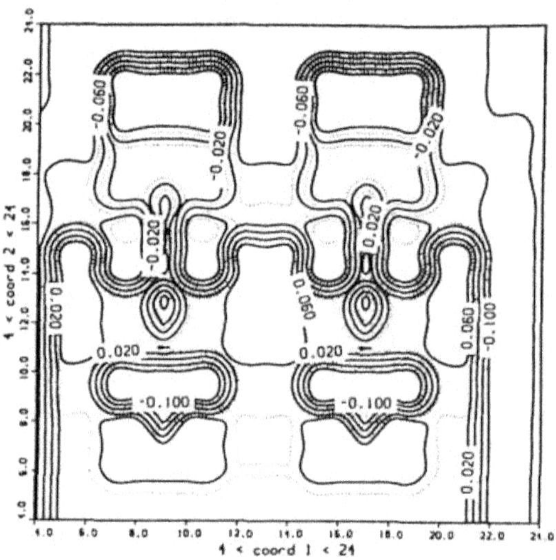

Figure 6.23:

5×5 **W4** $\eta = 5, \lambda = 0.5, a_0 = 2.8, a_f = 4.8, k = 9$, the spots are progressively stronger (a increasing) as one proceeds down the figure. The cross-coupling resistance is weak and the weakest spot, at the top, vanishes. The remaining eight spots are bounded into two groups.

Figure 6.24:

5×5 **W4** $\eta = 5, \lambda = 1.1, a_0 = 2.8, a_f = 4.8, k = 9$, the effect of increasing λ is marked; the actual location of the spots is almost lost in the diffusion bounds induced.

In the diagram of Table 6.6, the implicit lines, which the subject has to provide on seeing the display, are the linear joins of the pairs (W,X), (Y,Z), (A,B) and (C,D) *and only those pairs*. By the usual processes of perceptual grouping, if the figure is presented unconstrained, without instructions, then lines such as (Y,D), or (X,Z) may have a higher probability of being seen as reflecting natural Gestalten within the dots.

To put this point in a way which makes better sense in field theoretical terms; the intermediate instruction to the subject "tell me if the lines which join (A,B) and (C,D) are collinear[9] if extended towards filling the gap between B and C" needs some operations by the subject to create the necessary and sufficient virtual lines (A,B,C'), (B',C,D), (W,X) and (Y,Z), where the Poggendorf illusion is now equivalent to an assertion that the pairs B,B' and C,C' are not identities.

We have shown, rather piecemeal in $(n \times n)\Gamma^k$, that lines can be created between pixels and between points, and that points can be grouped by diffusion boundaries into local clusters, and that angles can be filled. Also, new collinear lines which echo short input segments can be induced, and symmetry properties of a total figure enhanced in the surrounding ground. There are regions in the *dots plus inplicit lines* figure, called here the pseudo-Poggendorf figure, which correspond to line junctions. The question then is, if the Γ process runs until points are joined up by diffusion to create contours, will those contours then next, in a second internal stage of cascade step sequences, take on the same role as the line inputs mapping directly onto pixels which serve in the original representation of the illusion figures?

As shown in Table 6.6, the groupings (A,B,X) and (Y,C,D) can be made so close that mutual interference within those subgroups occurs within the ηk trajectories before anything else. But if new points (w,x,y,z) are added as shown, then the phenomenology locally can change, and in our nonlinear Γ network dynamics quite different couplings will emerge.

Consequently, the question of what minimal dot configuration is needed cannot be answered without knowing the topography of the figure as well as its topology, and the perceptual processes involved in pseudo-Poggendorf are not necessarily the same as in Poggendorf, even if the eventual judgment biasses are analogous considered in terms of their outcomes as a quantifiable illusion.

---

[9] The perception of collinearity as a basic factor in perception was originally emphasised by Wertheimer (1923).

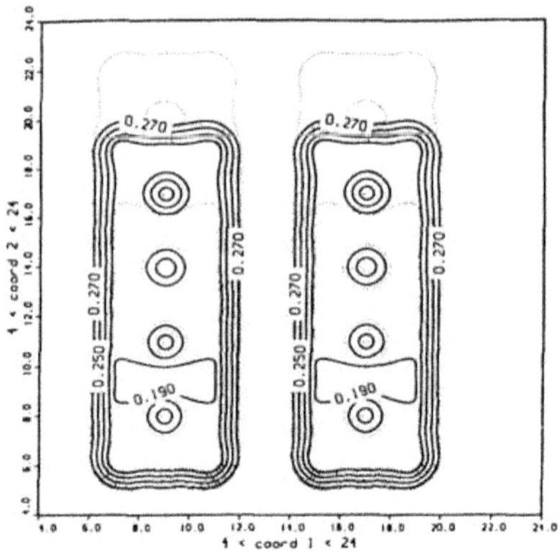

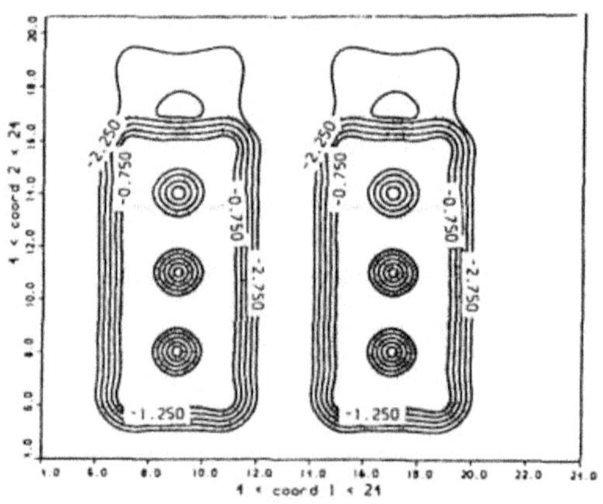

Figure 6.25:

5×5 **W4** $\eta = 5, \lambda = 0.5, a_0 = 2.8, a_f = 4.8, k = 11$, this entropy plot shows the persistence of the whole ten spot images, but at varying strengths, even with low λ.

Figure 6.26:

5×5 **W4** $\eta = 5, \lambda = 0.9, a_0 = 2.8, a_f = 4.8, k = 11$, *cross-entropy plot* for the same conditions as Figure 6.25; the weakest spot has vanished except for distorting contours. This effect is important because it operates as a threshold mechanism. the threshold is nowhere represented in the system as a specific constant, but emerges as a consequence of the nonlinear interactions between a_f and a_0.

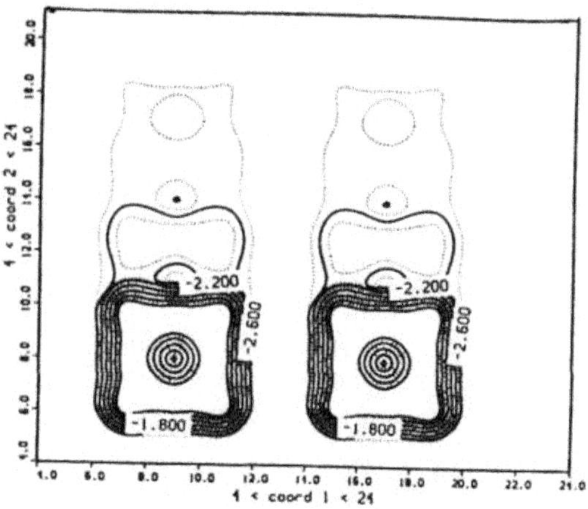

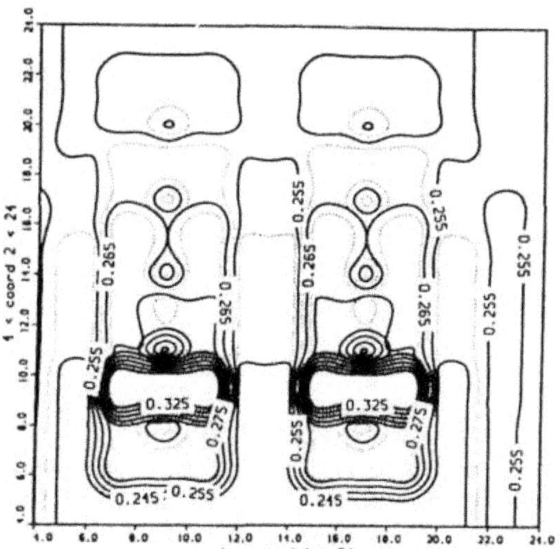

Figure 6.27:

5 × 5 **W4** $\eta = 5, \lambda = 1.0, a_0 = 2.8, a_f = 4.8, k = 11$, cross-entropy plot; this is almost the same situation as the previous case, but for a slight increase in λ. The gradient of output intensity is more pronounced.

Figure 6.28:

5 × 5 **W4** $\eta = 5, \lambda = 1.1, a_0 = 2.8, a_f = 4.8, k = 11$, entropy figure showing all ten points are present with complicated local interactions.

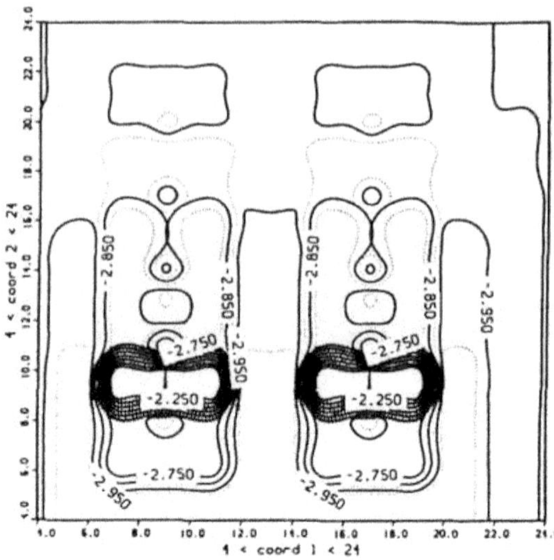

Figure 6.29:

5×5 **W4** $\eta = 5, \lambda = 1.1, a_0 = 2.8, a_f = 4.8, k = 11$, cross-entropy plot; again a slight increase in λ from the situation in Figure 6.27 induces changes. All ten spots are now revealed, the interaction between the strongest inputs is marked.

Multiple Perception and Multiple Process Representation

The significance of the curious Figure 6.1c is now examined. Rubin originally described our perception of this and other geometrically curious figures in his 1915 text *Synsoplevede Figurer* (in Danish), but the most accessible source is the later 1950 reference, in which alternative theoretical interpretations are reviewed. The point is, that if we attend to the segments with Müller-Lyer fins added, those two segments seem to be unequal in length, in the way which is usual for that illusion. At the same time the horizontal midline is seen as being parallel to the base, which is physically exactly what it is. There are thus two incompatible percepts, incompatible from the standpoint of Euclidean geometry. But they are not incompatible phenomenologically. The whole situation is unstable; Rubin summarises it (1950, p. 382,383):

> "There must be a certain complicatedness in a figure which can be used for demonstrating disagreement with geometry."

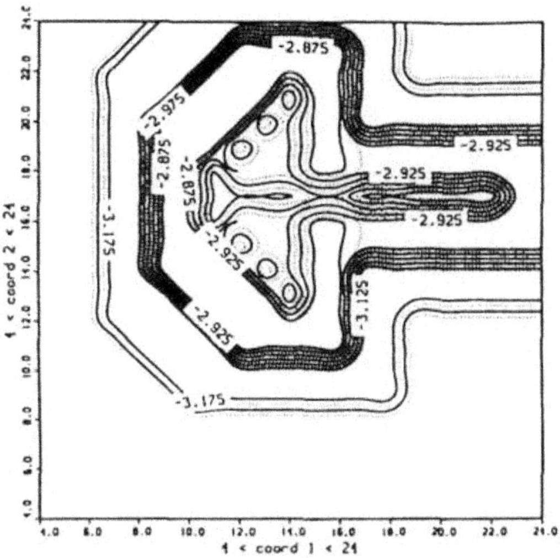

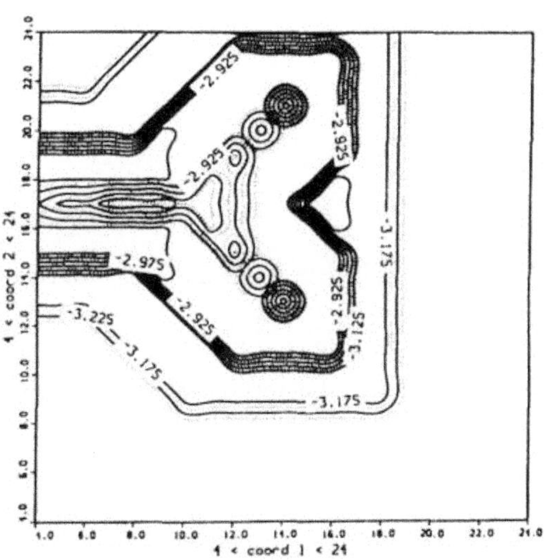

Figures 6.30 and 6.31:
Müller-Lyer arrowhead regions; after $\lambda = 1.2, \eta = 5, k = 9$. The shaft has not been plotted in its middle.

Table 6.6; the skeletal pseudo-Poggendorf figure.

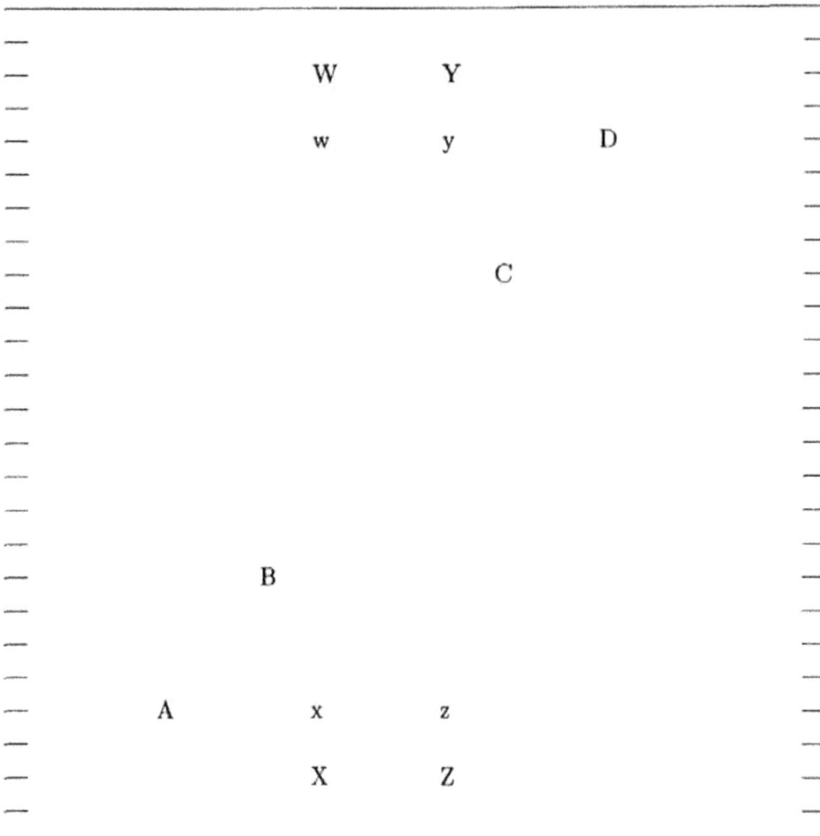

> "..with the Müller-Lyer figure there can be sensorially experienced a certain looseness or fluttering characterising the visual figure."

Rubin further noted that on sustained inspection figures lost, to some degree, the character of reality, and that the observer can have visual experiences of two features of a pattern at the same time; he wrote of 'a certain doubleness' in perception, one 'mental' and the other 'physiological' (meaning veridical) processes, and that the feeling of unreality was a result of the discrepancy between the two processes. Which, if any, of these multiple representations, and the fluttering, can be predicted as necessary consequences of tapping into the $(n \times n)\Gamma^k$ evolution, either at one or more points in the ηk sequence, or by integrating over a subsequence of the evolution, is addressed by inspecting some of the plots in chapter 5 in conjunction with the Müller-Lyer diagram in its complete form. There are two demonstrated properties of an $(n \times n)\Gamma^k$ field that can be relevant here; (a) the network will preserve the parallelism of line segments which do not intersect with other segments and which are collinear with the lattice axes, and (b) the network will induce filling and distortions in the neighbourhoods of line junctions.

Hence, in NPD theory, if the observer on judging parallelism uses predominantly information from line segments which are phenomenologically not extended into the local neighbourhoods of junctions (though they are parts of lines which physically do so extend), then the Rubin figure's perceived shape can be supported by networks, given also the fluctuation of attention from one local area to another which Rubin noted. When the viewer is looking focally at the horizontal lines within Figure 6.1c process (a) obtains, when attention is on comparing the two Müller-Lyer oblique segments then the fins induce (b) processes. There is no need to invoke two mechanisms, psychological and physical, as Rubin did, one sort of psychophysical transformation and constraint to local attention of the stimulus display are necessary and sufficient system properties.

Discussion

An interim conclusion is that using 3×3 or 5×5 $(n \times n)\Gamma^k$ rectilinear lattices makes no significant difference to the qualitative dynamics; using 5×5 makes it easier to see what is happening. The cascading of lattices induces both smearing and a marked instability in the levels of the pixels adjacent to the figure, at least within concave regions; this can be shown more readily by pushing the parameter values up to near the chaotic region, as in the Appendix to this chapter. The behaviour of the $(n \times n)\Gamma^k$ system can be loosely described as dynamically partitioning the field into three sorts of regions: (i) Where the figure of a_f pixels is induced, there is stability in position, though the level of the $a_{f,k}$ can fluctuate above and below the surround. (ii) Inside a concave bounded region, such as an acute angle most readily exemplifies, instability of level and configuration is induced, and a local region which can be above or below the $a_{f,k}$ is induced. This region may, or may not, be joined to the surrounding ground in the field, depending on the parameters λ, k, c_2. (iii) Outside the figure ridges and troughs are propagated, resembling Mach bands.

The propagation is abruptly terminated with increasing k if $c_2 \leq 1.0$, which therefore constitutes a difference between **W2** and **W4** for the 5×5 models.

The Poggendorf illusion is only about acute angles, but the Müller-Lyer figure has been produced in a diversity of shapes, the end angle terminators $<, >$ have been replaced with [,] or with circles ○, ⊖ and it is reported (Robinson, 1972) that the illusions are still produced. Using $(n \times n)\Gamma^k$ it is easier to fill an acute angle than to fill a [figure, because the two right angles have to be brought close together before the filling arises. A single right angle or ⊢ figure produces very little angle filling. It has been known for some time that the expansion form of the Müller-Lyer figure produces greater distortion that does the shrinkage form (Binet, 1895, Bayer and Pressey, 1972), and using entropy mapping it is readily found that the configuration near the shrinking arrow heads is quite different from the expansion region; there is a pinching or weakening of the shaft inside the shrinking arrow head, and an emphasis of the shaft locally behind the expanded arrow head. Both arrow heads have some internal filling, but to different degrees (Figures 6.30, 6.31). It is interesting, therefore, that Day and Dickinson (1976, p. 147) wrote "In brief, by attaching one or more lines to another figure spatial extents greater or or less than the test line are generated. The apparent size of the test line is perceptually assimilated to these extents". The conjecture here would be that some process called "perceptual assimilation" produces an output based on the process averages that the entropy plots approximate; at the final stage of a response as a conscious experience the patterns of activity within $(n \times n)\Gamma^k$ are resolved into one somewhat unstable image.

The theory here is not specifically about vision, and indeed from the work of Over (1968) who showed, following Révész, that haptic analogues of visual geometric illusions could be produced, it can only meaningfully be interpreted as mapping the perception of spatially extended physical inputs onto internal lattices which preserve some field properties inherent in the inputs. There have been various previously advanced field theories about visual illusion generation, and Robinson (1972, chap. 6) reviews them; none have the mathematical structure of $(n \times n)\Gamma^k$ and most are simply verbalisations of ideas of lateral inhibition or of gradients of influence across distances, of which Robinson was properly critical. One cannot offer any general criticisms of field theories because what properties a field has are infinitely variable, depending on both its algebraic representation and its parameter settings. The present approach not only involves nonlinear and complex functions across adjacent pixels and neighbourhoods, it evolves through time in a deterministic manner. and its parameter settings imply that contrast levels and rates of diffusion are also involved. There are experimentally testable consequences of such a structure, but the way in which many studies have been reported so far is insufficiently detailed to confirm or to refute the NPD approach.

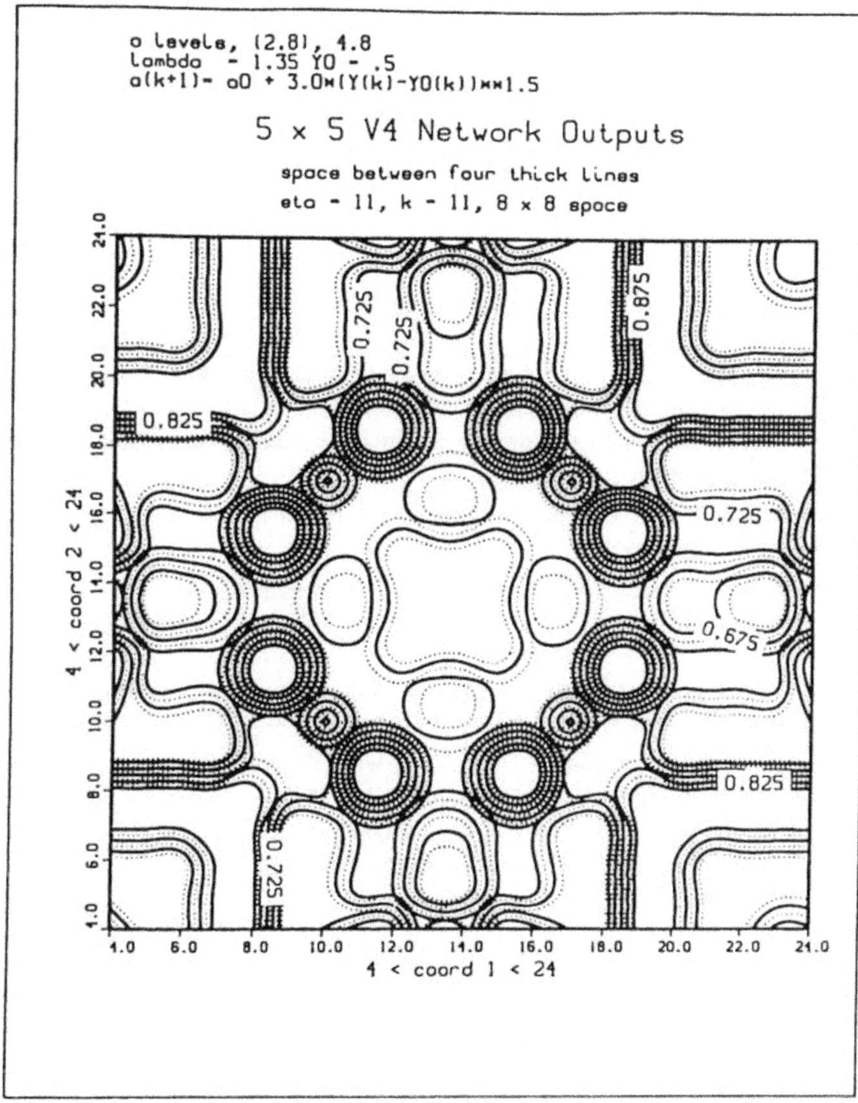

Conclusions

(1) Robinson's (1972) conclusion that field theories had not been illuminating in explaining the generation of geometrical illusions was true at the time.

(2) Many field theories involve reciprocal inhibition between points in a neighbourhood, but there is an infinity of possible mathematical models with such postulated properties. Without knowing the algebra, one cannot say what a field will do to its inputs.

(3) The field theory used here was not developed for investigating geometrical illusions; it is a general bivariate form of nonlinear psychophysics, NOT a model of any special sensory channel and NOT a model of the physiology of the retina of the human eye.

(4) The field theory used here is quantitative, deterministic, dynamic in that it extends through time, and irreversible in time. It has these properties precisely because it is an extension of nonlinear psychophysical dynamics, not because it is a field theory.

(5) Biological processes operate in a narrow parameter window; hence only some parameter settings of a field theory should be plausible.

(6) It is possible to derive, with appropriate parameter settings, envelope contours, angle filling, local image suppression, contour continuation, the induction of orthogonal contours, and in the limit chaotic image perturbations, complete ganzfeld effects, and local uncertainty of images. Obviously not all these arise at the same time.

(7) To compare nonlinear field theory with actual experiments requires precise settings of: figure-ground contrast, scale, image orientation, stimulus exposure times, response latency measurement, and subjective confidence of image identification point-wise over a subset of the field.

(8) One cannot properly test a field theory by breaking the input pattern into separable parts, and arguing only from responses to those isolated parts.

Notes

(1) Mach did not directly apply his mathematical models of contour perception to generate visual illusions based on line figures.

(2) A field theory advanced by Pressey and Pressey (1992) uses some of

the terminology of field theory, but the algebra is not in fact field-theoretic, but rather fitting ad hoc equations to some outcome effects.

(3) Helson (1933) listed 114 points of Gestalt theory; some of these can be given a sufficiently precise interpretation to be considered for their putative relevance to a field theory. Clauses in question, with Helson's reference numbers, are listed below. Our critical comments are in square brackets.

1. Absolutely homogeneous fields are not configurations.

2. A configuration is any segregated whole. [Can a lattice be equated with segregated ?]

4. A configuration is a co-existence of phenomena in which each member carries every other. [only if neighbourhood diffusion is extended through cascades]

5. To be a configuration a totality must be demarcated from the rest of the field. [$a_f \neq a_o$]

6. Configurations are units arising against more diffuse backgrounds. This identifies the configuration more with figure than with ground.

9. Configurations of all kinds represent dynamic processes in interaction.

10. Configurations are not sums of parts or of parts and relations.

15. Configurations are transposable. [no absolute coordinates in the lattice?]

19. Configurations possessing poorer structure are poorer configurations and lack properties such as clearness and persistence.

25. Certain properties apply only to configurational wholes; like slender, round, and symmetry.

30. Temporal configurations require time for their completion. Dynamic processes underly all configurations.

35. Configurational concepts do not represent absolutes.

38. Any process exhibiting direction in its changing states may be called a configuration.

39. Configurations are either dynamic or are the outcomes of dynamic processes which underlie them.

40. Static configurations are rarely met with; most configurations are either dynamic or quasi-stationary, their changes are too slow to be observed except over periods of time.

42. Configurational changes may be so slow as to escape attention or so rapid that their essential nature is overlooked.

44. Configurational changes unsually proceed from forces within the structure itself.

48. Configurations may change towards a more uniform level state in which differences tend to be obliterated.

52. Configurational changes occur in such a way that the time integral of energy expended is a minimum. [how could this be tested?]

71. Configurations cannot be analysed into... *any* optional parts and relations.

74. Configurations condition their parts, not parts wholes. [not quite right, for a network!!]

77. Maximum connectedness results in homogeneous totalities which are non-membered.

80. Configurations are responsible for connectedness; connectedness does not give rise to configurations.

98. Parts which are alike tend to go together to form groups. [circularity in defining 'like']

99. Parts having smaller intervals between them tend to form wholes rather than parts separated by larger intervals.

100. If several possibilities are present by which a part may be continued in a whole, the simpler more regular may be chosen.

109. Dynamical interaction, undisturbed by accidental impacts from without, leads to orderly distribution of forces and hence configurations. [this is the point where a classical field theory is invoked]

114. Learning consists really in discovery of configurational properties already present. [present in what?]

Appendix: Near Chaotic Dynamics

Sample matrices of a values show the evolution over k of the pattern inside a three-segment concave line figure, such as used on the ends of the Müller-Lyer figure in some versions.

The version used is **W4** for 25 inputs in 5x5 scanning over 27x27 lattice, only the centre part of the lattice shown, with parameters a_0, a_f, λ, c_2 and η: 2.800 4.800 1.350 1.500 10. The value of λ used here is near to that discussed in the chapter.

For $k = 3$

4.41	4.41	4.41	4.41	4.41	4.41	4.41	4.41	4.41	4.41	4.41	4.41
4.41	4.41	4.41	4.41	4.41	4.41	4.41	4.41	4.41	4.41	4.41	4.41
4.41	4.41	4.41	4.41	4.41	4.41	4.41	4.41	4.41	4.41	4.41	4.41
5.10	5.10	5.10	5.10	5.10	5.10	5.10	5.10	5.10	5.10	4.41	4.41
5.10	5.10	5.10	5.10	5.10	5.10	5.10	5.10	5.10	5.10	4.41	4.41
5.10	5.10	4.71	4.71	4.71	4.71	4.71	4.71	5.10	5.10	4.41	4.41
5.10	5.10	4.71	5.10	5.10	5.10	5.10	5.10	5.10	5.10	4.41	4.41
5.10	5.10	4.71	5.10	5.10	5.10	5.10	5.10	5.10	5.10	4.41	4.41
5.10	5.10	4.71	5.10	5.10	4.41	4.41	4.41	4.41	4.41	4.41	4.41
5.10	5.10	4.71	5.10	5.10	5.10	5.10	5.10	5.10	5.10	4.41	4.41
5.10	5.10	4.71	5.10	5.10	5.10	5.10	5.10	5.10	5.10	4.41	4.41
5.10	5.10	4.71	4.71	4.71	4.71	4.71	4.71	5.10	5.10	4.41	4.41
5.10	5.10	5.10	5.10	5.10	5.10	5.10	5.10	5.10	5.10	4.41	4.41
5.10	5.10	5.10	5.10	5.10	5.10	5.10	5.10	5.10	5.10	4.41	4.41
4.41	4.41	4.41	4.41	4.41	4.41	4.41	4.41	4.41	4.41	4.41	4.41
4.41	4.41	4.41	4.41	4.41	4.41	4.41	4.41	4.41	4.41	4.41	4.41
4.41	4.41	4.41	4.41	4.41	4.41	4.41	4.41	4.41	4.41	4.41	4.41

For $k = 13$

4.97	4.97	4.97	4.97	4.97	4.97	4.97	4.97	4.97	4.97	4.97	4.97
4.83	4.83	4.83	4.83	4.83	4.83	4.83	4.83	4.83	4.83	4.83	4.83
4.83	4.83	4.83	4.83	4.83	4.83	4.83	4.83	4.83	4.83	4.83	4.83
4.89	4.89	4.89	4.89	4.89	4.89	4.89	4.89	4.89	4.89	4.83	4.83
4.89	4.89	4.89	4.89	4.89	4.89	4.89	4.89	4.89	4.89	4.83	4.83
4.79	4.79	4.86	4.86	4.86	4.86	4.86	4.86	4.89	4.89	4.83	4.83
4.79	4.79	4.86	4.64	4.64	4.64	4.64	4.64	4.76	4.76	4.83	4.83
4.79	4.79	4.86	4.64	4.64	4.64	4.64	4.64	4.76	4.76	4.83	4.83
4.79	4.79	4.86	4.64	4.64	4.92	4.92	4.92	5.10	5.10	4.83	4.83
4.79	4.79	4.86	4.64	4.64	4.64	4.64	4.64	4.76	4.76	4.83	4.83
4.79	4.79	4.86	4.64	4.64	4.64	4.64	4.64	4.76	4.76	4.83	4.83
4.79	4.79	4.86	4.86	4.86	4.86	4.86	4.86	4.89	4.89	4.83	4.83
4.89	4.89	4.89	4.89	4.89	4.89	4.89	4.89	4.89	4.89	4.83	4.83
4.89	4.89	4.89	4.89	4.89	4.89	4.89	4.89	4.89	4.89	4.83	4.83
4.83	4.83	4.83	4.83	4.83	4.83	4.83	4.83	4.83	4.83	4.83	4.83
4.83	4.83	4.83	4.83	4.83	4.83	4.83	4.83	4.83	4.83	4.83	4.83
4.97	4.97	4.97	4.97	4.97	4.97	4.97	4.97	4.97	4.97	4.97	4.97

7 Other Field and Cascade Representations: Isosimilarities, and Instabilities

In this chapter some further applications of cascades and fields, with rather different properties but starting from the same seed ideas, are examined. The sense in which similarities are related to fields is not the same as the perceptual fields examined in the previous two chapters; instead the fields are an abstraction to represent the mapping from stimulus properties onto a contour map where the dependent variable is a relationship $\mathcal{S}(V, W)$ between points and not an output Y at a point. The psychometrics of similarity has produced a wide variety of theories, which have been reviewed (Gregson, 1975); it may be noted that none of the metric space models which have been pursued in multidimensional scaling are used here. This analysis relates to the treatment of separability in chapter 3; some definitions are repeated for convenience and their details expanded.

Abrupt jumps in long sequences of psychophysical responses have been known for a long time; they are often treated as a nuisance phenomenon and experiments are run until behaviour "settles down" and yields simpler data. Here such jumps are treated as something to be modelled because jumps in the phase space are expected in nonlinear dynamical processes, particularly ones which in time series analyses are modelled as multistate or Markovian. Some examples have been given already in the evolution of figure and ground pixels in chapter 5. There are various ways in which dynamic instability may translate into discontinuities in a system's output sequences, one case in audition is treated here.

Similarities

Similarity responses are made, in their simplest form, to pairs of stimuli. Each of the two stimuli in the pair has at least two components, on dimensions which are common to both stimuli. We do not here consider stimulus comparisons where similarity judgments are elicited and the stimuli have one and only one physical dimension in common upon which they are compared. Let components on the n dimensions be $u_1, u_2, u_3, ..., u_n$ and the stimuli V, W. Then

$$V = \{u_{v1}, u_{v2}, u_{v3}\}, W = \{u_{w1}, u_{w2}, u_{w3}\}$$

is for example a sufficient specification for three-dimensional stimuli.

There are two main approaches (Gregson, 1975), metric and set theoretic, to modelling the similarity $\mathcal{S}(V, W), 0 \leq \mathcal{S}(V, W) \leq 1$, where $\mathcal{S}$ is a real positive function of its arguments. One approach, with variations, assumes that dissimilarity $\mathcal{D}$ is a function of the distance $D(V, W)$ in a Minkowski space with parameter p. Again, for $n = 3$, in such a representation we have

$$D(V, W) = \left[\sum_{i=1}^{3} \alpha_i |u_{vi} - u_{wi}|^p\right]^{1/p} \qquad [7.1a]$$

and

$$\mathcal{D}(V, W) = f(D(V, W)) \qquad [7.1b]$$

where f is always monotone increasing. Note that the weights α_i are not always written into the model at this stage. The cases $p = 1, 2, \infty$ are sometimes given special consideration, and a duality between the ranges $1 < p < 2$, and $2 < p < \infty$ exists in the case of 2-space. If it were desired explicitly to introduce an error variance into [7.1a], then either: (i) It can be added to the u_{vi}, u_{wi} as

$$u_{vi} + \varepsilon_{vi}, \quad u_{wi} + \varepsilon_{wi},$$

$$\varepsilon_v \sim N(0, \sigma_v), \quad \varepsilon_w \sim N(0, \sigma_w)$$

and $cov(\varepsilon_v, \varepsilon_w) \rightarrow 0$. The Gaussian assumption N is an ancillary assumption independent of [7.1a].

Or (ii) The addition of one single error variance is made to write

$$\mathcal{D}(V, W) = f(D(V, W)) + \epsilon, \quad \epsilon \sim N(0, \sigma_D) \qquad [7.1c]$$

The implicit assumption sometimes made is that $\mathcal{S} = 1 - \mathcal{D}$, where both $\mathcal{S}$ and $\mathcal{D}$ are suitably scaled to be bounded in (0,1). This can be true if $\mathcal{S}$ and $\mathcal{D}$ are taken as constructs in a formal mathematical model. To emphasise this point, we will henceforth write $\phi\mathcal{S}, \phi\mathcal{D}$ for definitions using physically measurable attributes and based on $\mathcal{S}(V, W)$ or $\mathcal{D}(V, W)$, and in contradistinction we will call observed human similarity judgments $\psi\mathcal{S}(V, W)$. Much of the endeavour in multidimensional scaling (mds) has been to retrieve $D(\psi V, \psi W)$

where $\psi V = \{\psi u_{v1}, \psi u_{v2}, \ldots\}$ (but of unknown n) and analogously for ψW, from $\psi\mathcal{D}$ or $\psi\mathcal{S}$ data only, but taken over the whole Cartesian product set $\{\mathbf{V} \times \mathbf{W}\}, V \in \mathbf{V}, W \in \mathbf{W}$. $\mathbf{V}$ may or may not be identical with $\mathbf{W}$; if $\mathbf{V} \neq \mathbf{W}$ then the $\psi\mathcal{D}$ data matrix is said to be off-diagonal. Here none of the mds assumptions are being used, and the comments just made are for distinguishing some key assumptions met in the work of Shepard (1974, 1991), Young (1970) and the *Psychometrika Monograph* approach of Lingoes and Roskam (1973), from what is being done here. We may note that the assumption $\psi\mathcal{S} = 1 - \psi\mathcal{D}$ is empirically false, as shown by Swedish work, which proceeded generally along different lines from the spatial assumptions (Gregson 1975, chapter 3). That is, if we want to elicit $\psi\mathcal{D}$ it must be done directly, and not by first eliciting $\psi\mathcal{S}$ and then converting it by subtracting it from a constant.

Two basic models are required, the first, for similarity (compare [3.8]) is

$$\phi\mathcal{S}(v, w) = \frac{\mathbf{m}\{\cap\}}{\mathbf{m}\{\cup\}} = \frac{\sum_{i=1}^{n} min(u_{vi}, u_{wi})}{\sum_{i=1}^{n} max(u_{vi}, u_{wi})} \qquad [7.2]$$

and the actual observed data are $\psi\mathcal{S}$. The u_{vi}, u_{wi} are stimulus components over predefined common dimensions.

This is a set theoretic model, it assumes that each physical attribute dimension is known and can be measured, and that the psychological information being used by the observer is covered by those measures. If the observer is in fact using stimulus properties which are not the physical attributes being measured by the experimenter, then [7.2] is likely to yield meaningless predictions. For example one can obtain $\psi\mathcal{S}$ similarites between pairs of landscape paintings, but what quantifiable pictorial properties are substitutable in [7.2] is not a priori decidable. in judging properties such as similarity, symmetry or complexity.

To be precise $\mathbf{m}(u_{vi})$ should be written for u_{vi} (and so for all v and w terms in [7.2]), distinguishing the measure $\mathbf{m}$ from the attribute u in each case. If Gaussian error variance were to be admitted in [7.2], then it could be written on the u, this gives error distributions which are functions of quotients of extreme values of random deviates, each of which are exponentially but not normally distributed (Gumbel, 1958, p 141). It follows that the introduction of the same sort of noise could yield different error variances in [7.1a] and [7.2], if we admit the Gaussian assumption.

The form of [7.2] follows equation [5.462.3] in Gregson (1975), and is used variously again by Gregson (1976,1980). It is to be distinguished from other vector or 'content' models introduced by Eisler and Ekman (1959), and set theory rather later by Tversky (1977). Unfortunately few comparative studies of the models exist; Gregson (1976) compared seven models including [7.2]. Gregson (1985) showed for one data set published by Broderson (1968) that [7.2] was preferable to some other alternatives of both spatial and set theoretic forms. The empirical differences in what is predicted are not in most cases great, and the choice is to be made from considerations of structural and parametric parsimony.

Disputes over the applicability of metric space models to similarity psychometrics perseverate, sometimes without a wide consideration of the available evidence. Krumhansl (1978) and Nosofsky (1992) reviewed mostly North American research, Ashby and Lee (1991) compared basically two sorts of models, one derived from signal detection theory and one from mds in Euclidean space, with Gaussian noise. Nosofsky and Smith (1992) disputed the implications of Ashby and Lee's work and wished apparently to retain metric space similarities at least in some contexts. None of these studies, and others cited therein, actually gave maps of isosimilarity contours; but some maps of scaled spaces starting from the full paired-comparisons data set for $\psi\mathcal{S}$ for mds, are given. These are in fact not helpful, and it is not possible adequately to reconstruct contours from their data in the required form.

It is crucial to remark that none of the models referred to, including [7.2], of themselves make any quantitative predictions about the variance of responses, though they may a posteriori treat the residual errors as Gaussian, though this problem has now been addressed in $n\Gamma$ without residual noise (Gregson, 1994). In set theoretic models, again mean or geometric mean responses are used, and the residual response variance is taken as Gaussian; this leads into linear regression estimation of $\hat{\psi}\mathcal{S}$ on $\phi\mathcal{S}$, or of $\hat{\psi}\mathcal{D}$ on $\phi\mathcal{D}$. We use the circumflex to refer to a fitted mean. The goodness of fit is usually expressed in terms of the regression, and does not of itself take into account shrinking (or stretching) of the ranges of the $\psi\mathcal{S}$ on $\phi\mathcal{S}$. The assumption that error is normally distributed is now known to be equivalent to assuming that a high-dimensional attractor generates the *one* $\mapsto$ *many* results. This is not necessarily true, and here we proceed without the assumption but in effect use a low-dimensional attractor which generates its own output variability.

The second basic model is simply $n\Gamma$ as a precursor transformation as input to [7.2]; the two processes are in effect cascaded; $n\Gamma$, and then $\mathcal{S}$ operating on Y instead of on U. For each n-dimensional stimulus an $n\Gamma$ transformation is applied to the inputs, to give a new output vector in complex variables Y. The real parts of these are then used as input to the set theoretic definition of similarity $\phi\mathcal{S}$. It is possible to apply any $\mathcal{S}$ model (including metrics) onto the output of $n\Gamma$ in this way, but because of results already known (Gregson, 1976,1980,1985) only the one case will be explored here.

$$\left\{ \begin{array}{c} V \mapsto n\Gamma(\{a_{vi}\}, \lambda, \eta) \\ W \mapsto n\Gamma(\{a_{wi}\}, \lambda, \eta) \end{array} \right\} \mapsto \mathcal{S} = \mathcal{S}_\Gamma(V, W) \qquad [7.3]$$

The similarities obtained if the $n\Gamma$ transformation is not imposed first are the $\phi\mathcal{S}$ of [7.2], and the actual observed data are $\psi\mathcal{S}$.

Isodistance Contours in Metric Spaces

From the Eisler and Lindman (1990) and the Tversky and Gati (1982) results we know that in general the $\psi\mathcal{S}$ isosimilarity contours are not metric in form (convex and circles for $\Re^2_2$); they can be confusingly like metric contours over restricted data subsets, but the $\mathcal{S}_\Gamma$ **Case 2** contours always have the necessary property of being mixed convex-concave.

There are in fact very few published attempts to map isosimilarity contours directly, the difficulty in obtaining enough data points to define the shapes of contours unambiguously is considerable. One important exception is the Eisler and Lindman (1990) study, which gives some individual plots; there are great intersubject differences, and none of the plots are readily reconcilable with metric space assumptions. Also, the shapes of the fitted contours do change with the degree of similarity, and can be convex, or locally concave. Instead, the actual depiction is circumvented in mds, and in other non-spatial models; linear regressions of predictions on theory, for sets of ordered similarities, are taken as adequate bases for accepting some theory.

If it is assumed that dissimilarities are monotonic on distances in an underlying psychological space, and also that that space can be represented as metric, then the shape of isodistance contours is the same as that found by mapping dissimilarity judgments into a metric space. The contours in $p = 1$ (City Block) space are squares with sides at 45° to the axes, those in $p = 2$ (Euclidean) space are circles. For $p = \infty$ (the Supremum metric) the contours are a square with sides parallel to the reference axes (Gregson, 1975, sects. 4.51, 4.52). See the checklist given below. The Euclidean case is familiar intuitively. So, if similarities are represented, through transformations which preserve order over all pairs in a product set of comparisons, in $p = 2$ two-dimensional space, the contours representing degrees of similarity to a reference stimulus will be a family of concentric circles centred on the coordinates of that stimulus.

Metric space contours have been illustrated in various texts; the cases $p = 1$, $p = 2$, and $p = \infty$ are usually treated, because their contours are most simply definable analytically.

In contradistinction, the contours for set theoretic similarity models, although readily derivable, are less well known; Eisler and Lindman (1990) found that some contours were in one segment Pascal's limaçon, and in another segment hyperbolae. These are always convex with respect to a reference point, but may have singularities.

The contours for [7.2] and hence for ϕS are cyclic quadrilaterials, almost the same in shape as for the City Block metric; they are almost preserved for high similarites when $n\Gamma$ trajectories are applied for some λ values. The shape remains the same even if the range of stimulus values is shifted. This ϕS representation is not used in the later analyses here. It follows that for some stimulus sets used as input it would be difficult to separate the predictions of set theoretic and metric space models; one would have to contrive sets with critical values of $\{V, W\}$ to maximise the post hoc probability of differentiating the theories. If we use the nonlinear Γ trajectories here then theoretical isosimilarity contours are computable. It turns out that their form is not fixed, and can be surprisingly intricate, depending on η, e in **Case 1** and on λ in **Case 2**. This may be aesthetically pleasing, but also raises problems for comparing real data with theory.

Tversky and Gati (1982) noted that isosimilarity contours "do not all have the same shape; rather, they seem to be convex in the small even when they are non-convex in the large". They point out that in [7.1a] $p < 1$, e.g. $p = .67$, produces concave contours. This property they use as further evidence of the untenability of metric space representations, other properties of similarities suggest that even if variable p as a function of S is admitted, one cannot salvage the idea of metric psychological space. Unfortuately Tversky and Gati did not publish actual contour maps, but the diversity of the stimulus material they used, including pictures of simplified plants in pots, faces with eye and mouth changes, histograms with various modes and tails, suggests some generality to their comment. There are other difficulties with interpreting the Tverski and Gati study, in that sometimes they were concerned with relative similarities in triples of stimuli, which is not the same problem (Gregson, 1975, Chapter 1). A checklist, which is not exhaustive but also notes where ideas seem to have originated, internationally, follows.

Checklist of models of similarity used in the relevant source literature from about 1960 to the present
with indication of predicted forms of isosimilarity contours

Metric Spaces			
City Block	Minkowski $p = 1$	Squares at 45^0	USA, BRD
Euclidean	Minkowski $p = 2$	Circles	USA, BRD
Supremum	Minkowski $p = \infty$	Squares	USA, BRD
[intermediate cases, $p \neq 1, p \neq 2$ exist]			
Vector			
	Separated qualitative and quantitative	approximately circles	Sweden
Polymorphous			
	No single linear mapping	unknown (indeterminate)	Netherlands and Austria
Set theoretic			
	Intersection, Union and Ratios	convex with singularities	
	Eisler-Ekman	(sometimes	Sweden
	Gregson	confusable	Australasian
	Tversky	with metric)	USA
Nonlinear Psychophysics			
	$n\Gamma$ **Case 1**	convex-concave	Australian
	$n\Gamma$ **Case 2**	not invariant	

There exist other treatments which do not attempt to fit specific algebraic models, and so make no predictions about the shapes of isosimilarity mappings. This includes some work from Lithuania. Some precise source references are given in Gregson (1994).

Isosimilarity Contour Simulations

We now explore the three-way agreements between $\phi\mathcal{S}$, $\psi\mathcal{S}$, and $\mathcal{S}_\Gamma$ for a given set of data where the physical stimulus components are all known *a priori*. We do this by generating isosimilarity response surfaces; those derived from Γ and $n\Gamma$ are a large family; Figures 7.1 through 7.6 show examples. We have used a 3Γ process, but with a_3 always fixed at the minimum value of 2.001. This case is called $2-(3)\Gamma$. It is desirable to avoid a strictly even-dimensionality situation, because for the Euclidean special case distances are invariant under rotation, but not for $p = 1$. This property has been given some importance by Melara and Day (1992). No such property exists for $\mathcal{S}_\Gamma$ isosimilarities. There are two problems to be addressed in constructing simulations:

(i) to generate a collection of $\phi\mathcal{S}$ (a) and $\psi\mathcal{S}$ ($Y(\text{Re})$) which will span a reasonable range of values between 0 and 1,

(ii) to investigate the differences between **Case 1** and **Case 2**, particularly with variations in λ in the second.

Table 7.1: Key to Figures

Parameters of $2-(3)\Gamma$ used to generate each contour map.

$a_3 = 2.001$ throughout all maps.

Figure	**Case**	e or λ	η	a range
7.1	1	$e = .2$	10	$2.8 < a < 5.3$
7.2	1	$e = .2$	11	$2.8 < a < 5.3$
7.3	2	$\lambda = .5$	10	$2.1 < a < 4.6$
7.4	2	$\lambda = .6$	10	$3.2 < a < 5.2$
7.5	2	$\lambda = .9$	10	$3.2 < a < 5.2$
7.6	2	$\lambda = 1.1$	10	$3.2 < a < 5.2$

The contour maps are generated on a 25×25 matrix, each $\mathcal{S}_\Gamma$ value is related to the Y components at the reference point (the centrepoint in most cases) of the graph; at coordinates (13,13). The range of a values (that is, a_1 and a_2) that can be depicted without explosion of the Γ trajectories is restricted by the choice of e and λ. A wide range of a has been pictured for each of Figures 7.1 through 7.6. Apparently twisted contours are sometimes generated locally by the DISSPLA algorithm when contours are tightly packed in the space; in such regions they are better treated as imprecise. The examples generated are chosen to show some of the variety of $\mathcal{S}_\Gamma$ maps that can arise from the $2-(3)\Gamma$ nonlinear system with changes only in the second parameter, e, or λ when cross-coupling is introduced in **Case 2**.

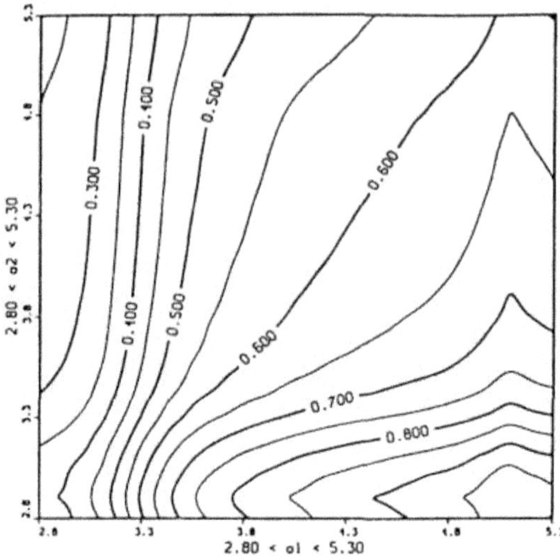

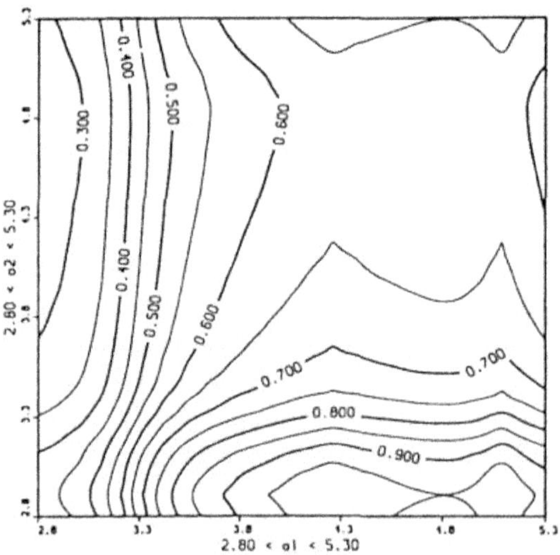

Figures 7.1 and 7.2 $2-(3)\Gamma$ with Case 1

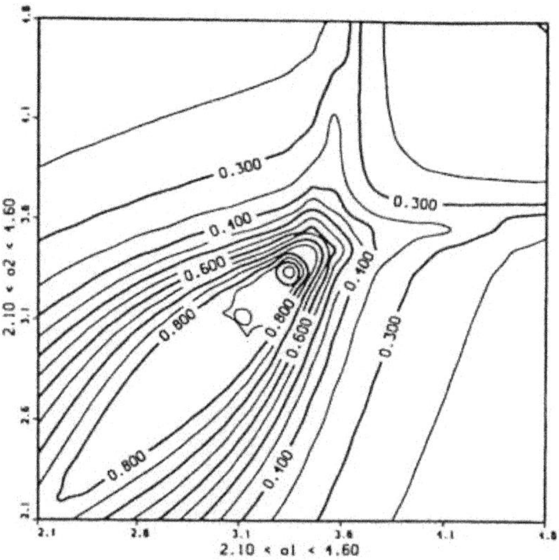

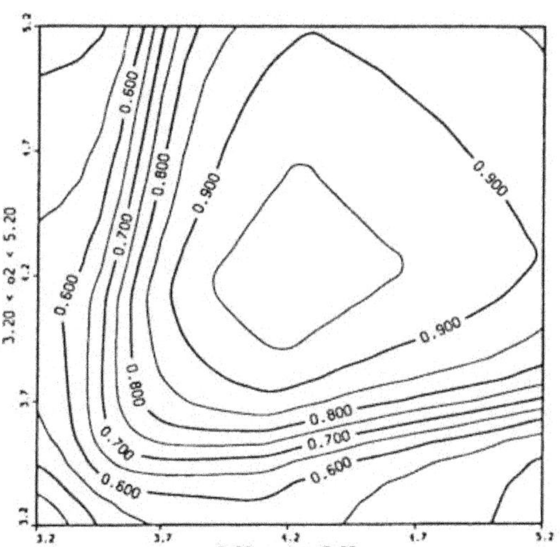

Figures 7.3 and 7.4 $2-(3)\Gamma$ with Case 2

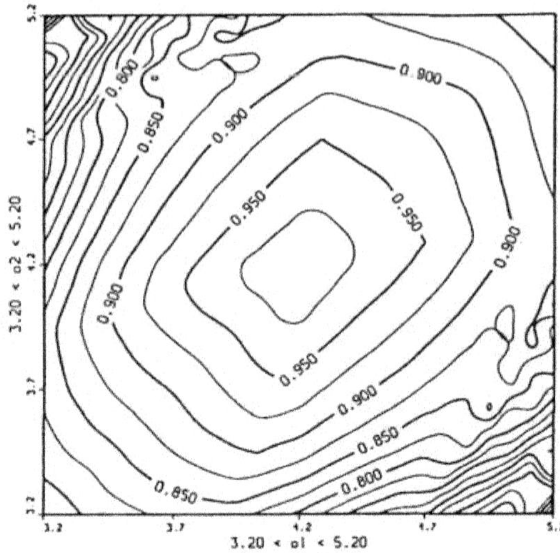

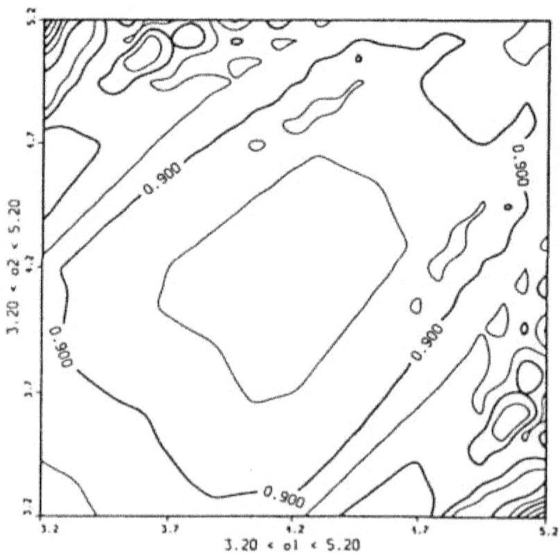

Figures 7.5 and 7.6 $2-(3)\Gamma$ with Case 2

It is useful, because only the metric space contour forms are commonly used or implied in published studies, to compare features of the figures both with $p = 1$ and $p = 2$ Minkowski contours, and with the Eisler and Lindman (1990) and Gregson (1975) illustrations. The most striking difference between metric space contours and the one produced here is that the families of metric contours are always the same, for example in $p = 2$ (Euclidean) space the isodistance and derived isosimilarity contours are a family of concentric circles. In $p = 1$ space they are a set of squares with a common centre point; the diagonals of the squares (parallel to the reference axes) are common. In all the Minkowski cases, usually $p = 1, 2, \infty$ are considered, the contours are convex with respect to their common centre point. In the $p = 2$ case they have no singularities, in the other two cases they do have them. Also, the metric space isodistance contours have the same shape irrespective of the coordinates of the reference point; one can move about and still have the same form of map, given that the axes are defined on a fixed linear scale unit. But in the maps produced here the shape of the contours varies with the degree of similarity, and can be convex or concave locally. It also varies according to where we take the reference point, putting it on the diagonal $a_1 = a_2$ does not yield the same contour map as taking it off the diagonal.

Each of the Figures 7.1 through 7.6 represents $\mathcal{S}_\Gamma$ relative to the centre-point of the plot, where by definition there is a single point with $\mathcal{S}$ and $\mathcal{S}_\Gamma$ equal to unity. It is the only identity point, where identity signifies $\mathcal{S} = 1$, in the a_1, a_2 parameter space. This is true no matter where the reference point is located; the gradients of the similarity surface fall away from the identity point, but not invariantly so. Exceptions locally can be created; troughs and secondary plateaux can exist. All of Figures 7.3 through 7.6 using **Case 2** must be symmetrical about the bottom left to top right diagonal with $a_1 = a_2$, a_3 fixed, until the system becomes locally chaotic. This is the only symmetry with reflection, and there are none with rotation. In this feature they obviously differ from the metric space contours which are symmetrical in a rotational group with period $\pi/2$, and never by definition exhibit chaotic dynamics.

Case 1: Dimensions not Cross-coupled

As **Case 1** appears to fit psychophysical data in fewer contexts than **Case 2**, (Gregson and Britton, 1990), only two examples, in Figures 7.1 and 7.2, have been chosen for illustration. The second parameter e is fixed at 0.2, and $\eta = 10$ (for even), and $\eta = 11$ (for odd iterations of the Γ trajectory) are used because high a components induce cycling in Y. The two figures are almost identical in, and only in, the lower left quadrant. The differences are marked in the upper right quadrant, exactly as expected from the nonlinear dynamics.

The irregularity of the isosimilarity contour shapes implies that they could resemble those of $p = 1$ in some regions and those of $p = 2$ in others. Misidentification of data as falsely being in metric space could therefore readily occur with data that are noisy and unevenly distributed over the parameter space. For **Case 1** the range of $\mathcal{S}$ values is too narrow; it is difficult to generate very low similarities, and they are associated with comparisons with the coordinate origin.

The interest in Figure 7.2 lies in the upper right area; now four separable regions of $\mathcal{S}_\Gamma > .95$ develop, thus violating the monotonicity of gradients away from the centrepoint. Combining this with Figure 7.1 suggests a prediction of response instability when one of the 2 – (3) stimuli has very high component values. The rhombus-like contours in the middle of both Figures 7.1 and 7.2 resemble those of $p = \infty$. In Figure 7.1 the $\mathcal{S}_\Gamma = .9$ contour has a slight resemblance to a contour derived by Eisler and Ekman (1959).

Case 2: **Dimensions Cross-coupled via λ**

In this series, for $\eta = 10$, Figures 3 through 6 show the effects of increasing λ, over a range of λ values used in previous studies already cited.

The changes in the contour map with λ are extensive. There might be an apparent paradox lurking here, in that for a given stimulus set (the a parameter space with regard to the reference point) when λ is low, representing weak component resistance to masking by other components, the range of $\mathcal{S}$ values is greater. As λ increases, so that dY/da is not mutually inhibited, the range of $\mathcal{S}_\Gamma$ is constrained upwards, except in small regions where the contours are close packed, and there appear indications of chaotic dynamics.

As λ increases the gradients adjacent to the central reference point flatten. The contours become more generally convex; they approach a shape which is like a rectangle with some curvilinear distortion. In practice data generated by $n\Gamma$ **Case 2** would be readily misidentified using multidimensional scaling algorithms as in $p = 1$ or $p = 2$ space according to where the empirical data were most dense. Again, the changes due to η are mainly away from the lower left quadrant; there is evidence of chaotic dynamics when $\lambda > 1$ and $a > 4.6$ on one dimension. If $a > 4.6$ on both dimensions, the contours are oddly entwined but the gradients are not as steep as in the top left and bottom right corners of Figure 7.6. If the reference point were to be moved to the bottom right corner, this would not produce a simple displacement of the contour map, but a change in its form. The closed contours apparently vanish in the middle region. Further cases are given by Gregson and Mahar (1994), for various types of geometric figures.

Transitions in Contour Forms

It is obvious that some contours are locally nearly like segments of circles, others nearly bits of straight lines, and that as $\mathcal{S}$ increases concavity turns into convexity. All the plots here are dependent for their detailed chracteristics on the DISSPLA interpolations of the 25×25 data lattice. Granted that approximation to an ideal mapping under perfect parameter continuity, there are still major determining variables of the $n\Gamma$ onto $\mathcal{S}$ process that are identifiable; namely the range of a values used, and the cross-coupling via λ. In terms of these system parameters, there is an interesting 'saddle-point' in the dynamics, for $\mathcal{S}_\Gamma$ around 0.5. An ancillary way to describe the phenomenon is to plot the frequency distribution of $\mathcal{S}$ values over the 625 data points, for each plot. In some cases, this derived probability distribution becomes effectively trimodal in $\mathcal{S}$. One may also therefore compute the information transmitted by each frequency distribution, and note that it will be low in those circumstances where the trimodality appears.

Table 7.2: Information in Maps

$2-(3)\Gamma$ **Case 2**

a ranges 3.2 to 5.2

λ	H	H/H_{max}
0.300	3.305	0.7647
0.400	3.380	0.7819
0.500	3.318	0.7676
0.600	3.191	0.7383
0.700	3.023	0.6994
0.800	2.794	0.6465
0.900	2.474	0.5724
1.000	2.326	0.5382

Identification

The depiction of isosimiliarity contours given here is radically different in its derivation from previous treatments; most invariances are lost, but at the same time the topography of the $\mathcal{S}_\Gamma$ surfaces is defined over all the stimulus parameter spaces.

The distinction between **Case 1** and **Case 2** is not only of interest within $n\Gamma$ theory, it has analogues in the issue concerning the separability of perceptual dimensions (Nosofsky, 1985, 1986, Shepard, 1991) already examined in chapter 3. If dimensions do not interact, in their information transmission from peripheral receptors to higher levels (in Γ theory this is the trajectory $U \mapsto a \mapsto Y$) then **Case 1** should be more appropriate as a prediction basis for similarity generation. If, however, dimensions are rarely fully separable, and interact in varying degree, **Case 2** is *a priori* a better choice. In the present context, the choice is not between separable and inseparable transmissions, but of something between, with complete separation a rare event. In mds the separability condition would be represented by $p = 1$. In separability studies (for example, Melara and Marks, 1990, Melara and Day, 1992, Nosofsky, 1985, 1986) the rate at which classification tasks are performed is a confounding variable, under time pressure $p = 2$ may be more appropriate, whereas similarity scaling studies *per se* rarely introduce this additional complication.

The two fundamental questions which motivated this study were really, (i) can we get at similarity within a body of psychophysical theory that has some demonstrated generality outside the specific problems of similarity representation, and (ii) are there sufficient invariant properties of isosimilarity contour maps which support their identification from real data ?

To the first question the answer is affirmative, metric space assumptions are not psychologically necessary even if they are statistically attractive. But the second requires for its pursuit the use of statistical methods which are unlike those exploitable if the analytically simple assumptions of metric space properties are retained.

In any real experiment we cannot hope (under stationarity) to cover the S surfaces like Figures 7.1 through 7.6 in anything like the density which enables us to determine gradients which are steep or change rapidly. The most that can be done is to put an open lattice of data points onto the continuous theoretical surface, and match the two at the lattice points while assuming functional continuity between them. This treatment of sparse data is possible with maximum entropy methods of chapter 2. It was noted earlier that choosing between similarity models could be difficult, because if we use linear regression of data on theory as a measure of goodness of fit, contending available models can predict much the same ordering of similarities, and various theories all fit data fairly well; they appear to take up over 80 percent of variance. But comparing models by assessing how well they fit isosimilarity contour maps is quite a different proposition. We are now concerned with the matching, pairwise, of the topologies of similarity maps treated as response surfaces, and have only sparse actual data spaced on lattices on those surfaces.

From nonlinear psychophysics a diversity of response surfaces can be created; Table 7.3 lists some; these are all precursors to using [7.2] which would then create similarity contour surfaces from pairs of points on corresponding response surfaces.

In practice the problem of matching theory to sparse data is one of a class of such problems in image analysis, and uses Bayesian inference with very non-standard priors. This topic has received a lot of attention; a symposium (Journal of the Royal Statistical Society, Series B, 1993, *55*, p 3-102) on Markov Chain Monte Carlo and Gibbs Sampler methods is a pertinent example.

The response surfaces we may construct from $n\Gamma$ theory are continuously definable mappings of nonlinear functional relations, not only is the output predictable, but also some components of the associated variance and the local response latency distributions. What is not predictable without strong ancillary assumptions is the numerical scaling of responses; the surface topology is the focus of interest here. There is, in Bayesian statistics, no absolute measure of goodness of fit; we are concerned always with choosing between alternative models, given a minimum degree of parametrisation, and prior credibility of the models. The $Y(\mathrm{Re})$ output of Γ always lies in the interval (0,1) unless it has exploded (which circumstances are not considered in the psychophysics of normal behaviour) and in $n\Gamma$ within $0 \leq Y_{tot}(Re) \leq n$, which without loss of generality is rescaleable to lie in (0,1). The response units of ψS are arbitrary and usually on a bounded response scale; by definition

$$0 \leq \phi S \leq 1 \Leftrightarrow complete\ difference \leq \phi S \leq identity$$

Any measure on the interval (0,1) over the system's parameter surface can be taken as a basis for the minimum cross-entropy and this given us an index of the relative proximity of data to theory. It is not merely that this approach is possible; in some sense it is optimal. The models are acting as reference priors (Hills, 1987). It is not worth doing this with the metric space contours. because a *priori* they are known to be wrong. A related problem in identification arises if the observed contours are treated as deformed images in the S space. In Minkowsi space the set of deformed images always contains the *pure* image. If the deformations are an i.i.d. sequence, which in terms of experimental psychology means repeated noisy independent trials, then the resultant image

Table 7.3: Response Surfaces
which are definable from $n\Gamma$ theory

(1) One input and one output; variable sensitivity

$$Y(Re, Im) = \Gamma(a, e)$$

for a family of psychometric functions.

(2) Derivatives of (1) for indentifying regions of maximum entropy and predicted instability

$$\nabla Y = \Gamma(a, e)$$

(3) Total output from two inputs without cross-coupling of the e parameters

$$Y_1(Re) + Y_2(Re) = \Gamma_1(a_1, e_1) + \Gamma_2(a_2, e_2)$$

denoted by **Case 1** in Gregson (1992a)

(4) Total output from two inputs cross-coupled via λ

$$Y_{tot} = 2\Gamma(a_1, a_2|\lambda)$$

Comparison of (3) and (4) estimates some forms of hypoaddition.

(5) Generalisation to n dimensions

$$Y_{tot} = n\Gamma(a_1, a_2, ..., a_n|\lambda)$$

$$Y_{tot})Re) = \sum_{k=1}^{n} Y_k(Re)$$

where the $Y_k(Re, Im)$ are individual component outputs; because of the cross-coupling

$$\forall k \; : \; Y_k = f(a_1, a - 2, ..., a_k, ..., a_n|\lambda)$$

This situation needs hypersurface representation.

(6) A restricted case; n fixed equal multiple inputs; variable λ

$$Y_{tot} = f(a, \lambda, n)$$

(7) Derivatives, as for (2), exist for (3) through (6).

will tend to a circle, which is not the original image (Grenander, 1993, pp. 513-514, theorem 12.1.2) This means that averaging over large data samples does not clean up images, but actually distorts them in such a way that weak $n\mathcal{S}_\Gamma$ data will be misidentified as in $\Re_2^m$, $m < n$. This point, and the convex-concave shapes of isosimilarity contours, are relevant to showing that a theory of Melara (1992) that similarities are in a mixed metric of $\Re_1^m$ and $\Re_2^m$ is fundamentally wrong; the idea of a mixture of contours floating between two extremes which are everywhere convex, and in some cases but not all have singularities, is mathematically incoherent and empirically false. Melara's claim rests on a review of data which mostly omits cases contrary to his theory.

An Experiment with Three-Component figures

This experiment is concerned with visual inspection of line drawings whose geometrical properties are completely specifiable in elementary terms [1].

Method

Using three-column histogram figures, in pairs, presented visually so that the column heights reading from left to right are $u_{v1}, u_{v2}, u_{v3}, u_{w1}, u_{w2}, u_{w3}$, a constant sum display was created with the sum of the heights of the columns set at 19 units, and the reference stimulus located at (5,5,7). Corresponding pie slice diagrams were also constructed; the pie slice representations are alternative components instead of columns. The plane in which the similarities are mapped is $1 \leq u_1 \leq, 1 \leq u_2 \leq 9$, the hidden plane is then $1 \leq u_3 \leq 17$. The theoretical isosimilarity contour map, with $a_i = c_1 u_i + c_2$ is generated in 3Γ for various λ in **Case 2**.

Four conditions are used in this group of experiments; they are designed to exemplify the two forms of $2-(3)$ constraint we have used, and to display the stimuli in bar and pie form. For $2-(3)$ stimuli with a constant $\mathbf{m}(u_i)$ the mapping from bar heights to pie segment angles is $1 \Leftrightarrow 1$, because of the constant sum and the same metric $\mathbf{m}(.)$ for all components in each display. But when one component bar is fixed in its physical height, in our case $\mathbf{m}(u_3) =$ *const.*, then the mapping onto the circle (composed of pie slices) has to be proportionate within each stimulus, and the correspondence breaks down. The effect of this is one feature we wish to investigate.

---

[1] Some parts of this experiment were previously reported in the proceedings of the Palma, Majorca conference of the International Society for Psychophysics in 1993.

For brevity, the abbreviations

$$B_{CS} = \text{bars with constant sum of heights}$$
$$P_{CS} = \text{pies with constant sum of angles}$$
$$B_{FX} = \text{bars with one fixed component in height}$$
$$P_{FX} = \text{pies with proportions of angles matching } B_{FX} \text{ heights,}$$

will be used hereafter.

Nine different subjects participated indiividually in each condition. They were all volunteers from the Psychology Department at the Australian National University, and naïve. about NPD and similarity psychometrics. Each subject covered 81 lattice points in a random sequence.

Results

Considering the pooled (coordinate-wise over 9 Ss) maps in Figures 7.7 to 7.10 for the four conditions, each generated on a 9×9 data lattice,

(1) All pooling induces some smoothing; some local crossing of contours is due to the DISSPLA software having to cope with close points.

(2) The associated maps of the s.d.s (not shown) for each of the 80 points $(9 \times 9 - 1)$ are heteroscedastic; irregularly increasing in variance away from the central reference point. the variance should increase with distance, though that is an oversimplification (Gregson, 1994).

(3) The maps for B_{CS} and B_{FX} are simpler, and the mean similarity for B_{FX} is lower from inspection of the figures.

(4) The P_{CS}, B_{CS} transformation is apparently one rotated through $\pi/2$ on a virtual axis of the contour ridge.

(5) The B_{FX} is simplest and in its central region could readily be mapped erroneously into metric space.

(6) The P_{FX}, which is in either $\Re^3_2$ or in $\mathcal{S}|3\Gamma$, is the most tortured in configuration; the most interesting feature is the high ridge running up to the top right corner of the map. In the $\cdot_{FX}$ cases the proportionate mapping necessitated by pie picturing induces a quite different isosimilarity contour topology. This has a psychological interpretation, namely that $\psi\mathcal{S}$ judgments in the B_{FX} condition are not on the proportional value of components, the third bar is hence of diminished importance as it has no associated information (in the uncertainty sense), and should have no significance in $\Re^3_1$ or in $\Re^2_1$.

The results of Table 7.4 confirm that the different pictorial representations of the same u_1, u_2 with fixed or compensatory u_3 can induce different judgment distributions; there is no invariance from u_1, u_2 to $\psi\mathcal{S}(U, V)$, contrary to the

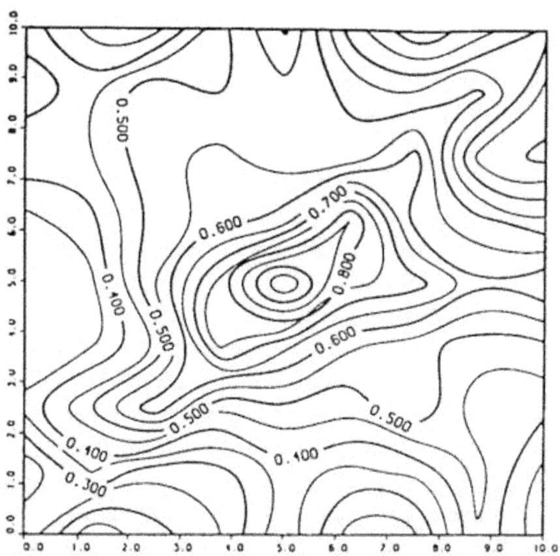

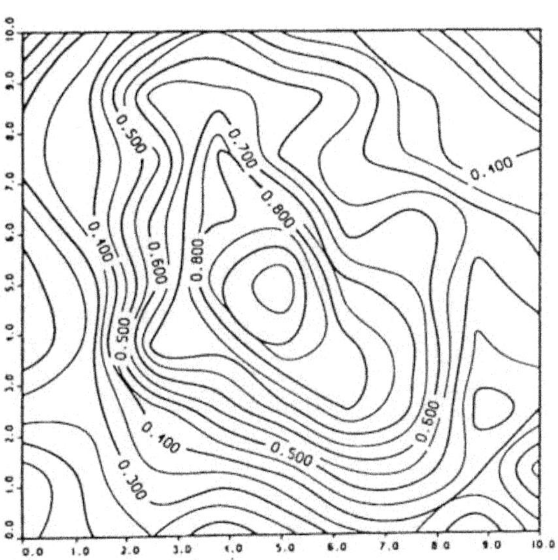

Figures 7.7 and 7.8 conditions B_{CS}, and P_{CS}.

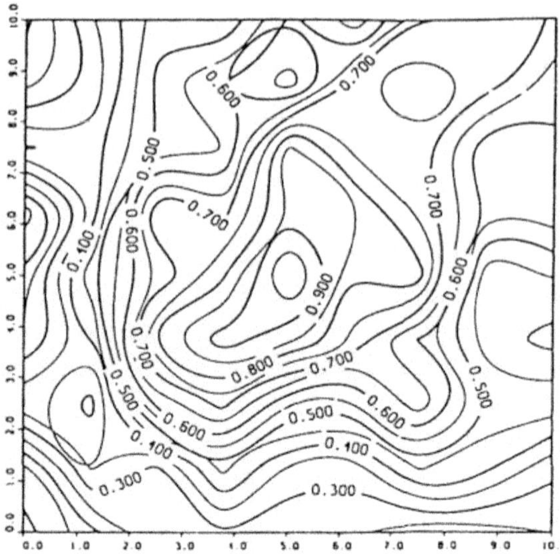

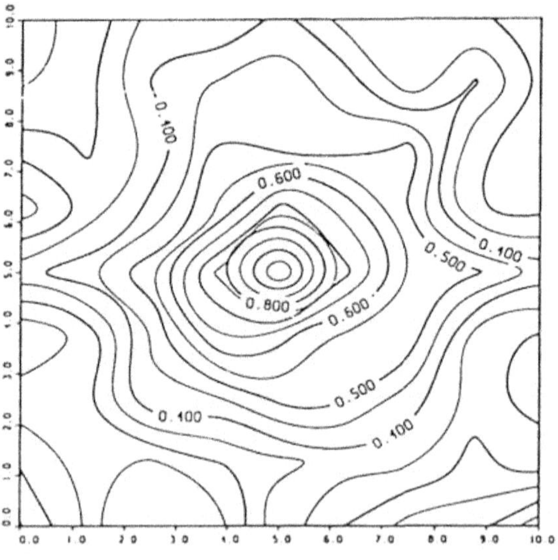

Figures 7.9 and 7.10 conditions B_{FX}, and P_{FX}.

Table 7.4: Comparison on Similarity Response Distributions
Using Kolmogorov-Smirnov Two Sample Tests
(cells are the test statistic d(.))

	B_{CS}	B_{FX}	P_{CS}
B_{FX}	.081		
P_{CS}	.115	.158	
P_{FX}	.156	.199	.077

From Walsh (1965, p. 98) d(.01), ($N_1 = N_2$=720), = .1062
Mean $\psi\mathcal{S}$ values: B_{CS}=.416, B_{FX}=.396, P_{CS}=.473, P_{FX}=.514

predictions of metric space representation.

Discussion

There is a beguiling simplicity about the metric space idea with its regular convex contours as the outcome; it is unfortunately very wrong when we treat the problem not as one of scaling but as one where the psychophysics involves indentifiable stimulus and response properties.

Not one of the 36 individual isosimilarity contour maps we have derived (not illustrated), nor any of the four pooled mean maps (Figures 7.7-7.10) are simply convex; none resemble the isodistance contours of metric spaces. The four variance maps associated with the four pooled mean similarity maps are all heteroscedastic. That is, they are not planar, and all show an irregular tendency to increasing variance away from the centre reference point. In that sense, therefore, psychophysical maps of human similarity judgments made to visually presented stimuli are definitely incompatible with the metric space assumptions inbuilt into some multidimensional scaling studies. The metric space assumptions are wrong, and in any case are unnecessary for psychophysical investigations even though appeals to mathematical tractability at the expense of psychological validity persist.

The question of interest then, is whether the labile but persistently convex-concave contours we obtain, which persist in that form with smoothing through coordinate-wise averaging, can be matched by any psychophysical theory. Further insight into how $n-(n+k)\Gamma$ processes create the convex-concave contours we observe can be gained by increasing the number of fixed components to two; then we have $\mathcal{S}|2-(4)\Gamma$ and two situations to be distinguished. If the $n-(n+k)\Gamma$ structure is extended to 4 dimensions, then the two fixed components, which we may call a_3 and a_4, can be set constant in two interesting ways. Either we can treat the situation as one with two low level noise terms,

$a_3 \simeq a_4 << \{a_1 \neq a_2\}$, which is where we have redundant background noise, or alternatively we can trap the variation between two anchor levels, so that $a_3 << \{a_1 \neq a_2\} << a_4$. This has interesting consequences, the condition with two low (almost fixed) noise components produces isosimilarity contour maps in a_1, a_2 space which resemble the data for the condition of $2-(3)$ bar figures with constant sum B_{CS}.

But the trapping condition produces contour maps which resemble more the very distorted non-metric contours reported by Eisler and Lindman, to which they tried to fit Minkowski $p = -\infty$, (not a metric space by definition) and to which Tversky and Gati reported $p \simeq .76$. Theoretical plots are shown in Figures 7.11-7.14 for various values of λ.

The implications are that the metric space approach will not distinguish these two $2-(4)$ situations, fixed u_i make no contribution to [7.2] no matter if they are minimal or trapping values, whereas nonlinear psychophysics for $n\Gamma$**Case 2** has to distinguish them.

We conclude that human similarity judgments are invalid as a starting point for scaling *globally* in metric spaces; it may be possible to scale them *very locally* in the neighbourhood of the systems' set point, and mathematical topologies which are only locally metric do exist but have not been used in psychophysics; for example Sierpinski (1956) refers to locally metric spaces defined in 1929 by Bieberbach, and Alò and Shapiro (1974) refer to locally convex linear spaces describable by Minkowski metrics. It would appear that the transformations of $\mathcal{S}_\Gamma$ used here map a globally **m**-metrically-definable stimulus space into a response space that can only locally resemble metric space, and this local space only has its metric topology preserved in the neighbourhood of the reference point. The evidence of Figures 7.7 to 7.10, and the previous studies cited, can be treated as the outcomes of nonlinear processes which have already been identified as plausibly underlying other psychophysical mappings not involving similarities *per se*.

General Comments

The treatment in this analysis has focussed deliberately on the form of isosimilarity contour maps. This is obviously only part of the story; the correspondence of similarity data, not necessarily in the shape of isosimilarity contours, with theory is also critical. Published results either convert sets of pairwise similarities into distances (D in hypothetical psychological spaces, via a monotone transformation $\psi\mathcal{S} \mapsto \mathcal{D} \mapsto D$ or they take observed mean similarities $\hat{\psi}\mathcal{S}$ and match them directly to theoretical similarities such as $\phi\mathcal{S}$ in [2a].

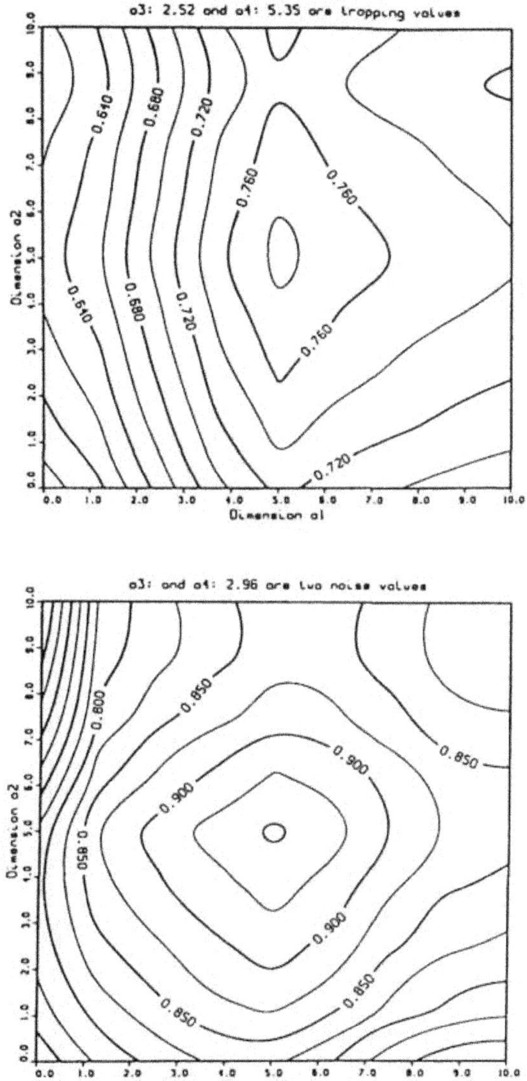

Figures 7.11 and 7.12

Theoretical maps for $\mathcal{S}|2-(4)\Gamma$, with trapping and with noise components

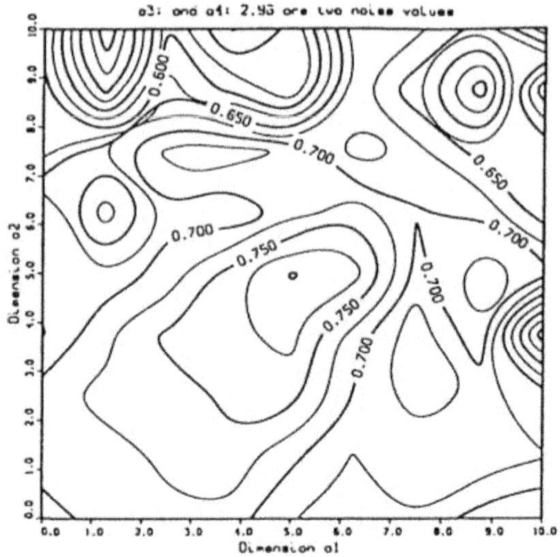

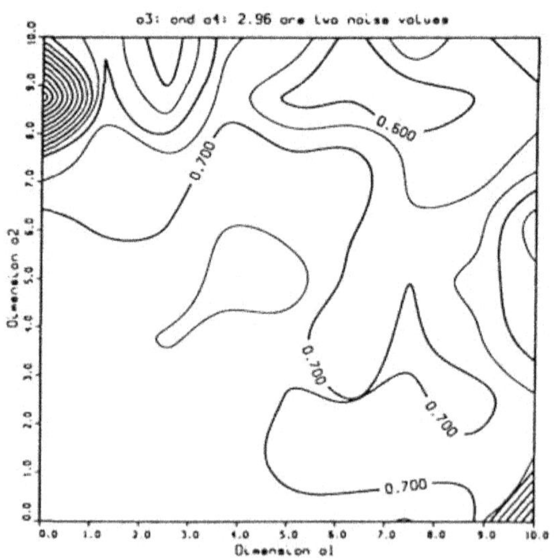

Figures 7.13 and 7.14
Theoretical maps for $\mathcal{S}|2-(4)\Gamma$, with noise components.

There were no published studies which matched $\psi\mathcal{S}$ to $\mathcal{S}_\Gamma$ directly.

The results cited at the start show that $\mathcal{S}$ based on [7.2] together with stimulus component measures u will match $\hat{\psi}\mathcal{S}$ as closely, or closer than, alternatives. It therefore raises the question, why make the transformation $\phi\mathcal{S} \Rightarrow \mathcal{S}_\Gamma$ at all ? If the isosimilarity contours of $\mathcal{S}_\Gamma$ differ substantially from those of $\phi\mathcal{S}$ and at the same time $\phi\mathcal{S}$ fits data well, then $\mathcal{S}_\Gamma$ must surely be worse. The answer to this rests on the role of λ in **Case 2**; in a parallel analysis (Gregson, 1993c) we have shown for some real data sets, where the $\hat{\psi}\mathcal{S}$ and $var\psi\mathcal{S}$ are known over a set $\{V \times W\}$ that the nearest to an identity transformation between $\phi\mathcal{S}$ and $\mathcal{S}_\Gamma$ arises, when using geometrical figure stimuli, as λ approaches 1.1. That value of λ is also one that appears optimum for other visually presented simple stimuli with physically definable properties. Figures 7.7 through 7.14 thus represent cases which may not equally probably arise in practice, even though Figure 7.9 bears some resemblance to Eisler and Lindman's (1990) Figure 1. Neither figures are compatible with metric space, both show concavity and convexity mixed, despite an assertion by Eisler (p. 168). The real value of Figures 7.3 through 7.6 is to show that the effect of a single system parameter, λ, is sufficient to induce the great lability of form in contour shapes. Because of this lability, locally they can match, or approximate to, some contours derived from alternative models of $\mathcal{S}$ with quite different algebraic rationales.

The Eisler-Ekman model is a case in point; possibly there is a range of very special stimulus conditions whould could apparently be constructed to support some of the diversity of similarity models that have been proposed. Nosofsky (1985) conjectured that very similar, that is difficult to discriminate, pairs induce Euclidean contours and readily discriminable pairs induce nearly City Block contours, which is an attempt to hang onto the psychological space assumption. Our example in Figure 7.3 weakly comes near to this idea. It would of course be possible to to put the $n\Gamma$ model in tandem with other set theoretic or matric models, in the two-stage construction employed here; the consequence would be contours which again would break down the invariances in shape with respect to $\mathcal{S}$ values which is exhibited in metric space or content models. There is an additional problem in the identification of any models that predict isosimilarity contours of consistently convex or concave shape for all $\mathcal{S}$. As pointed out by Gregson (1975), and reiterated in greater detail by Eisler (1989), two or more such similarity models can have identical, or nearly identical, contour shapes at least in some segments, and given the sparse coverage of real data over the contour surface, there is no practical way of establishing a likelihood ratio test that would differentiate between two such models. Eisler (1989) consequently called such models 'data-equivalent'; they predict exactly the same thing, up to the limits of resolution of real data. However, when we come to consider the $\mathcal{S}_\Gamma$ predictions, the situation is different because the shapes of contours vary considerably with the system's parameters, and collateral experiments to define those parameters (such as λ) are possible for different elicited judgments, other than similarity, on the same stimulus series. The nonlinear mappings via Γ trajectories are sufficient to break down these contour shape invariances, because the transformations are a function of at least two parameters, a, e or $a\lambda$ for each dimension.

The fact that $\psi\mathcal{S}$ similarity contours, when carefully plotted, do not resemble the metric space isodistance contours, no matter what value of p in the Minkowski equation [7.1a] is used, has led to an evasive argument by some writers committed to the notion of an underlying psychological space. Instead of trying to find a new model of such a space, which is not necessarily metric, if it exists at all, similarity itself is redefined as an unobservable, which supposedly then underlies the generation of various sorts of observable data. Observed $\psi\mathcal{S}$ data are then postulated to be in some way biassed by attention and memory weighting variables, which proliferate prodigiously; in a full model of k n-dimensional stimuli these parameters to be estimated could in number well reach $(n-1)(k-1)+(n-1)k(k-1)/2$. Nosofsky (1985, p. 415, 1992, p. 38) attempts to define the very troublesome $\psi\mathcal{S}$ out of the representation problem, by asserting that "the meaningfulness of defining similarity in terms of open-ended subjective judgments can be questioned", and "observed behaviour reflects only indirectly the underlying similarity representation". What is then scaled is not $\psi\mathcal{S}$ but the postulated $\phi\mathcal{S}$ which is a priori defined to be metric, and so is scaleable in that circular argument. In fact, here similarity as an operation is not *underlying* at all, but *overlying* an $n\Gamma$ mapping.

Arguments such as Nosofsky's we treat as ones of trying to define a difficult but essentially soluble problem out of existence, in order to cling onto a paradigm (in the Kuhnian sense) which is neither psychophysiologically plausible nor mathematically necessary for tractability.

Categorisation and Discriminants

The problem of allocating individual instances to category membership, using the properties of probability density maps in n-space has an extensive literature in mathematical and applied statistics, and in artificial intelligence. Given any point in the system's parameter space (an n-vector) it is possible, given some conditions on probability density functions, to construct likelihood functions for subclass membership, a posteriori; the discriminant function methods due to Fisher, and the Bayesian extensions are often to be found in standard computer software packages, and are used in taxonomic methods in various discipline such as zoology, archeology and clinical psycholology.

The now standard methods rest, as remarked, on being able to map probability contours over the parameter space; each subclass has a centroid and membership of a subclass is defined in terms of proximity to the centroid, where probability of subclass membership usually falls off monotonically with distance from the centroid. Probability contours are convex and with some transformation are usually ellipses, but it is sufficient that they are closed and definable.

Now replace probability contours by isosimilarity contours. Remind ourselves that similarities are bounded in (0,1) but do not satisfy the axioms of probability; attempts to say what axioms, if any, they do satisfy are inconclusive (Gregson, 1975). Intuitively one might wish to assert that the transformation from a probability hypersurface to a similarity hypersurface should be an isomorphism; we are not convinced that real human behaviour works that way. But, given what has been plotted, one could by methodological analogy with discriminant functions produce relative similarity plots, which would be like likelihood ratio plots. If metric space similarities exist, then the decision partition functions between subclasses should be straight lines, but if the relative similarities arise from the $n\Gamma$ surfaces which have some psychological validity, then these partitions could take on a diversity of very curious, but quite definable shapes. Computer scientists are much concerned to attack the problem of how we decide what we see is an orange and not a pear; using the similarity surfaces which have the very nonlinear contours described here could predict some real behaviour and distinguish that behaviour from optimum strategies in AI. It all depends on whether one wishes to describe similarity judgment behaviour, or to improve on it. We could do both.

Figures 7.15 and 7.16 give examples of relative similarity contours, both from $\phi\mathcal{S}$ using [7.2], which are locally straight lines, and after using the $2-(3)\Gamma$ transformations, which become very labile, both as a function of the direction of the line joining the centroids, and of λ. It is not surprising that linear discriminant function analysis can run into trouble if one attempts to disentangle systems of the sort we can get in nonlinear psychophysics.

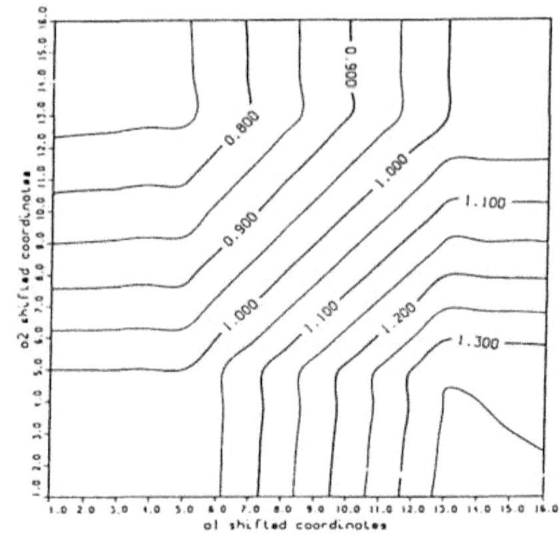

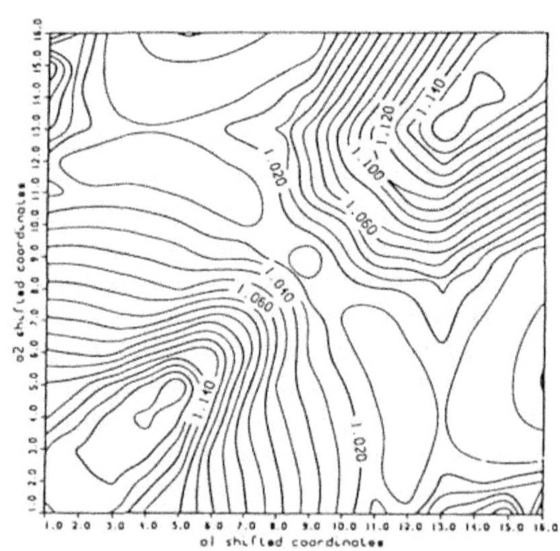

Figures 7.15 and 7.16

Relative similarity contour maps.

The Generation of Abrupt Shifts in Output from a Four Series Nonlinear Psychophysical Input

Auditory streaming is a well-documented psychophysical phenomenon, and has been the subject of a computer simulation network model by Beauvois and Meddis (1991). Given two physically-defined stimulus series simultaneously presented, what the observer reports as perceived is one of two patterns, each sustained for a while and then abruptly switched to the other. They alternate, or may eventually stick on one of them.

It is of interest to see if the generation of abrupt jumps between the perception of one and then of the other series, given that the input to the observer is a mixture of two simultaneous streams, follows from the dynamics of a 4Γ process under some restricted parameter settings.

Various studies (Anstis and Saida, 1985, Ciocca and Bregman, 1987, Bregman, 1978) have pointed to the long-term instability of the perception of auditory streams, and explanations of the observed switching between output characteristics have been attributed tentatively to higher-level decision processes (Bregman and Tougas, 1989), or to lower-level physiological transduction mechanisms and networks (Beavois and Meddis, 1991). Both Anstis and Saida (1985) and Beavois and Meddis (1991) present in diagrammatic form the system phase portrait, where regions of *streaming* (or *segregation* in alternative terminology) with an intermediate region of ambiguity, are delineated.

The published results suggest that the boundary between coherence and perceptual ambiguity is a first-order function of input frequencies, and a second-order function of the interstimulus interval of the tone series used, whereas the boundary between ambiguity and streaming (called the temporal coherence boundary) appears to be a first-order function both of frequency and the interstimulus interval. Possible jumping between outputs would, from these data, most readily be associated with small interstimulus intervals and lower frequencies. Clearly the total picture is one of nonlinearity, which motivates the construction of computer simulation models whose parameter settings are crucial.

The underlying psychophysiology has identified at least two pathways, located in the antero- and posteroventral cochlea nuclei, whose transmission properties are different; the anteroventral cells transmit the fine detail of signal inputs, whilst the posteroventral cells filter sequences and in effect smear detail. This multiple representation inplies that if we have two auditory stimulus series, then internally there will be four series to be processed and resolved to a single output at higher levels of analysis. This is the first point in this investigation where a parallel, though perhaps one too loose, can be drawn between the underlying dynamics of auditory streaming and those of a 2Γ network. The resemblance is too great to be passed over without consideration.

What we have to work towards here is the creation of a phase space for 2Γ, as a function of the parameter a (corresponding to frequency), and a new parameter ν (corresponding to intertone interval in real time), and the comparison of its fuzzy interphase boundaries with the empirical figures given by Anstis and Saida (1985) and by Beauvois and Meddis (1991).

Slipped 2-Gamma Structure

In the form of two recursions per dimension, 2Γ may be written in full (Campbell and Gregson, 1990) as two two-dimensional recursions in the reals x_1, y_1, x_2, y_2:

$$x_{1,j+1} = a_1(e^2 - e^2 x_{1,j} + x_{1,j}^2 - y_{1,j}^2 + 3x_{1,j}y_{1,j}^2 - x_{1,j}^3) \qquad [7.4a]$$
$$y_{1,j+1} = a_1 y_{1,j}(-e^2 + 2x_{1,j} - 3x_{1,j}^2 + y_{1,j}^2), \qquad [7.4b]$$
$$x_{2,j+1} = a_2(e^2 - e^2 x_{2,j} + x_{2,j}^2 - y_{2,j}^2 + 3x_{2,j}y_{2,j}^2 - x_{2,j}^3) \qquad [7.5a]$$
$$y_{2,j+1} = a_2 y_{2,j}(-e^2 + 2x_{2,j} - 3x_{2,j}^2 + y_{2,j}^2). \qquad [7.5b]$$

where $j = 1, 2, ..., \eta$, and boundary conditions are imposed on a_1, a_2, e (see Gregson, 1988, 1992). The recursions [7.4b], [7.5b] can show intermittency, and [7.4a] and [7.5a] generate the observable outputs between which switching could occur for high η. In [7.4a],[7.5a],[7.4b],[7.5b] we have set $e_1 = e_2 = e$ for simplification, but in 2Γ **Case 2** each $e_1 = f(\lambda, a_2^{-1})$ and $e_2 = f(\lambda, a_1^{-1})$ and the trajectories are cross-coupled through f, the coupling strength depending on the scaling parameter λ. We do not know, at this stage in the argument, if **Case 1** (no cross-coupling) or **Case 2** (cross-coupling) is most appropriate.

Introduce Γ as an operator notation (Gregson, 1992a Chapter 10), so that

$$Y_{j+1} = \Gamma_{a,e}(Y_j), \quad Y \; complex, \; j = 0, ..., \eta. \qquad [7.6]$$

It follows that

$$Y_\eta = \Gamma_{a,e}^{\eta}(Y_0) \qquad [7.7]$$

if the recursion is terminated abitrarily at η steps, and in the notation used in Gregson (1988),

$$Y_{obs} = (Re)[\Gamma_{a,e}^{\eta}(Y_0)] \qquad [7.8]$$

where Y_{obs} denotes the externally observable real part of Y after termination of the Γ trajectory. All of [7.6] to [7.8] are, of course, written for a single channel. If the termination of a recursion is not fixed but depends on its internal dynamics, then for completeness we should replace η with $f(J(a,e))$, which is a function of the non-constant Jacobian of the Γ trajectories. Under this condition Y_{obs} becomes a probability distribution over j and not a single value. In this treatment such complications are not fully considered, for simplicity.

Superimposition on a Single Dimension

The first variant of a Γ model with an input extended over an unbroken time subperiod of k steps within the time $0, ..., j, ..., \eta$ proceeds as follows.

The initial Γ recursion starts at t_0, with $Y_0 = (0.5, \epsilon_0)$, At $j = 1$,

$$Y_1 = \Gamma^1_{a,e}(0.5, \epsilon_0) = \Gamma_{a,e}(y_1, \epsilon_1) \qquad [7.9]$$

But the input is still persisting, and overwrites the ${}_1\Gamma^1$ with ${}_2\Gamma^0$ (slipped one j step later in its starting time), so that now

$$Y_{j+1} = {}_2\Gamma_{a,e}(max(0.5, y_1), max(\epsilon_0, \epsilon_1)) \qquad [7.10]$$

$$= {}_2\Gamma_{a,e}({}_2y_1, {}_2\epsilon_1) \qquad [7.11]$$

and the input still persists, and overwrites again by the same [7.10] rule, with

$$Y_{j+2} = {}_3\Gamma_{a,e}(max(0.5, {}_2y_1), max(\epsilon_0, {}_2\epsilon_2)) \qquad [7.12]$$

and so on up to $j = k$, when the last overwriting with the same (a, e) parameters occurs. After that period the process evolves from the dynamics initiated at

$$Y_{j+k} = {}_{k+1}\Gamma_{a,e}(max(0.5, {}_ky_1), max(\epsilon_0, {}_k\epsilon_1)). \qquad [7.13]$$

This presupposes that the input which persists is read into the Γ recursion at the same discrete time steps as the j.

The effective trajectory, with up to k superimposed trajectories, is given by [7.10] applied at each j step. If the input does not arrive at every j^{th} step, because the internal clock runs faster than the gating that lets the input in at discrete intervals, so that instead the input steps are at νj, where ν is an

Table 7.5: Components of Superimposed Recursion in One Dimension

		$j=0$	$j=1$	$j=2$	$j=3$	$j>3$
${}_1\Gamma$	=	${}_1\Gamma^1(Y_0)$	${}_1\Gamma^2(Y_1)$	${}_1\Gamma^3(Y_2)$	${}_1\Gamma^4(Y_3)$	....
${}_2\Gamma$	=		${}_2\Gamma^1(Y_0)$	${}_2\Gamma^2(Y_1)$	${}_2\Gamma^3(Y_2)$	....
${}_3\Gamma$	=			${}_3\Gamma^1(Y_0)$	${}_3\Gamma^2(Y_1)$	....
${}_4\Gamma$	=				${}_4\Gamma^1(Y_0)$	....

integer slippage coefficient, then the rows in Table 7.5 go in lateral jumps of ν and not in 1s as we read down the table.

This structure, which for brevity will be called $(k,\nu)\Gamma_{a,e,\eta}$ exhibits, in terms of Y_{obs}, rather uncomplicated dynamics until the (a,e) values push the Γ recursion past the first bifurcation. If $\nu j > \eta$ then we have no slippage, but instead *cascading* of the sort used in chapter 3.

Summation on n Parallel Slipped Inputs

Consider the implications of the notion that once a Γ recursion is underway, the channel in which it is active cannot receive further input until the recursion terminates. In physiological terminology, the trajectory is also a refractory period, blocking new inputs.[2] Obviously this idea is in contradiction to the previous superimposition notion.

However, there are $n-k$, $k \leq n-1$ empty channels in parallel waiting for inputs in an $n\Gamma$ system, and when one is occupied another can receive an identical $U \mapsto a$ input at ν lags after the last input. So the superposition operation [7.10] is replaced by **Case 2** cross-linkage via the scaling parameter λ and the observable output is now Y_{TOT}, a sum of the n $Y(\mathrm{Re})$ terms. Y_{TOT} and linear regression through a network were compared in chapter 1. Fitting one linear model to such a diversity of stimulus conditions as have been reported in auditory streaming studies would be transparently absurd and either uninformative or trivial in its statistical analysis. Fitting one nonlinear model with the same parameters would be a comparably inept exercise. What is needed, if a coherent psychophysical theory is to be advanced, is a careful consideration

---

[2] Once an individual Γ channel drops out, by whatever mechanisms are involved, (η or $J(a,e)$), it again becomes accessible if k is not by then passed and the input stimulus exhausted. Hence if k is much greater than $max(\eta)$ over n channels there will become possible some cycling of inputs over the n channels. We need to consider this situation because it is possible that $max(n)$ is as low as 6. The refractory period idea already mentioned in chapter 3 is given extended consideration in the Gamma Zak Embedding (Gregson, 1993a).

of which loci in the system's parameter space are implied by any one set of experimental conditions. Those conditions have to be specified both in terms of the stimulus input levels and intrinsic complexity, and the state of the organism in terms of its sensitivity and its prior exposure to analogous inputs. The choice, within subranges, of the (a, e, η, λ, n) parameter values has to be interpretable consistently in terms of what we already know or believe about the psychophysiology and the psychology of the sensory and perceptual processes implicated. Of course, what we know might not be enough, and what we believe can be wrong, particularly if inferred from a linear modelling exercise.

A Simulation using 4-Gamma

The empirical data suggest that a simulation should exhibit strong autoregression of outputs under specifiable conditions. These would include the use of two input levels which are fairly close, an output shift over a long series from predominantly one level to the other, and the presence of some system noise. The total output level should remain approximately constant throughout; it is the qualitative character of the output, and not its intensity, that exhibits the phenomenological jumps.

In any Γ model it is axiomatic that random (or high fractal dimensionality) noise can only arise in the environment of the organism, the internal trajectories are relatively low dimensional and deterministic. A series of inputs, where h is the counter variable on successive inputs, and k is the dimension label, with for example $k = 1, ..., 4$, and $h = 1, ..., 90$, are the stochastic series $a + \epsilon < h, k >$ and yield the output matrix $Y_{obs} < h, k >$.

The noise, therefore, is random perturbation on successive inputs a_h in one dimension, and the noise itself has an amplitude which is variable. For simulation, we use a rectangular distribution with zero mean, range $-.5, .5$, and a multiplier $\epsilon_{h,k}$ which either increases or decreases gradually with h. Independence of noise levels across dimensions is assumed.

For the results in Table 7.6, the $a_{0,k}i$, $k = 1, ..., 4$ values are set at 4.10, 4.10, 3.85, 3.85, with $\lambda = 1.35$, $\epsilon_{0,k} = 1.20$ for the descending noise series, and $\epsilon_{0,k} = 0.40$ for the ascending noise series. The increment/decrement in ϵ for one h step is .01.

This total configuration is intended to represent bilateral input, simulataneously to each ear, of a mixture of two tone sequences a_1, a_2, where each ear gets both mixed all the time. There are thus four input series, and each one has a complex Y representation in the internal Γ trajectories. So, for two input series, reduplicated and noise perturbed, we obtain a total of eight output time series. The cascading facility in Γ sets $Y_{0;h+1,k} = Y_{\eta;h,k}$ for all h, k. A property of Γ and hence of 4Γ is that the series of inputs has a zero autocorrelation spectrum (for all lags, being i.i.d.), whereas the output series Y_{obs} is autocorrelated, whether or not it is cascaded. It is to be expected that the lower order autocorrelations of the component $Y)obs < h, k >$ series would be higher in the cascaded process. The crosscorrelations $r(Y_{obs} < g, k >, Y_{obs} < h, k >,\ g \neq h$ are not readily predictable because they are a function of $\{a\}, \lambda, \epsilon$.

In Table 7.6 the code (k) of the a dimension with the greatest $Y(\mathrm{Re})$ is given for each of 90 successive trials; the lines run on as in text. The observed series are highly autoregressive, the distinctions between a_1, a_2 and between a_3, a_4 being lost in the coding, because we have set $a_1 = a_2$ and $a_3 = a_4$, though actual $a_{h,k}$ values are not equal as a consequence of the superimposed second-order noise ϵ. The code '1' inplies a higher input $a = 4.10$, and '2' a lower input $a = 3.85$ but the maximum output $Y_{h,k}(\mathrm{Re})$ varies from trial to trial.

Table 7.6: Sequences of Dominant Dimensions

90 inputs in series, 4Γ **Case 2** with Cascading

Example with progressively increasing noise

0 1 1 1 1 1 1 1 1 1 1 1 1 1 2 1 1 2 1 1 2 1 2 1 1 1 2 1 1 2
2 1 1 1 2 1 1 1 1 1 2 1 2 1 2 1 1 1 1 2 1 2 2 2 2 1 2 1 2 2
2 1 2 2 2 2 2 2 2 2 2 2 2 2 1 2 1 2 2 1 2 2 2 2 2 2 1 1 2 2

Example with progressively decreasing noise

0 1 2 2 2 1 1 2 2 1 2 2 2 1 2 2 1 2 2 2 2 1 2 1 1 1 2 2 2 2
2 1 1 1 2 1 1 1 1 2 1 1 2 1 2 1 1 1 1 2 1 1 2 1 1 1 1 1 2 2
2 1 1 2 2 1 1 1 1 1 2 1 1 1 1 1 1 2 1 1 1 1 1 1 1 2 1 1 1 1

Table 7.7 shows the values of the largest single dimensional output corresponding to the second example shown in Table 7.6, that with decreasing noise. It is assumed that the largest output is dominant under those situations where only one output is detected. Where a mixture is detected then a function of the total output is required. For this reason it is of interest to see how much the largest output deviates from the average over successive inputs.

The example in Table 7.6 resembles a situation where either $Y_1(Y_2)$ or $Y_3(Y_4)$ is heard, but never a mixture. The distinction between increasing and decreasing noise is of minor interest, so for heuristics we use only the decreasing noise example.

We next examine the three time series of Y_{obs}, Y_{max} and $Y_{TOT} - Y_{max} = \delta Y_{max}$ in parallel. The process can be regarded as a surface with Y_{TOT} as the

Table 7.7: Y**(Re) magnitudes of the Dominant Dimension**
90 inputs in series, 4Γ **Case 2** with Cascading
Example with progressively decreasing noise

0.0000	0.8996	0.7262	0.6914	0.6787	0.6968	0.7070	0.7066	0.7127
0.7156	0.6918	0.7071	0.7177	0.7166	0.7221	0.7077	0.7099	0.7152
0.6938	0.7048	0.7142	0.7035	0.7179	0.7177	0.7162	0.7172	0.7127
0.6937	0.6838	0.7043	0.7086	0.6895	0.7187	0.7120	0.7046	0.7149
0.7141	0.7024	0.7069	0.7110	0.7109	0.7051	0.6831	0.7092	0.7057
0.6989	0.7179	0.7162	0.7158	0.7062	0.7081	0.7140	0.7144	0.6898
0.7137	0.7170	0.6969	0.7110	0.7111	0.7027	0.7003	0.7033	0.6946
0.7099	0.7127	0.6984	0.7070	0.7018	0.7073	0.7091	0.7082	0.7028
0.7016	0.7103	0.7152	0.7089	0.7118	0.7120	0.7090	0.7090	0.7135
0.7046	0.7122	0.7025	0.7108	0.7030	0.7146	0.7132	0.7146	0.7133

first independent axis, λ as the second, and δY_{max} as the dependent variable. What we are then examining is the tendency of Y_{max} to dominate Y_{TOT} (and induce streaming) as a function of the Γ **Case 2** cross-coupling effect. As a_1 and a_2 as not very different, this is almost the same as varying e in Γ and seeing if the instability at the edge of streaming is associated with moving into limit cycling and chaotic dynamics within the trajectories of some or all $Y < h, k >$ for all h. Exploration of the output series with a range of $0.6 < \lambda < 1.5$ and $0.4 < \epsilon < 1.2$ suggests that an intricate interplay of the effects of these two parameters exists. With low noise the shift from a series of '1' to a series of '2' (as in Table 7.5) occurs fairly sharply around $\lambda = 1.40$. In other words, the dynamics are almost knife-edge with high cross-coupling, and lock into long repetitive output series of one sort of another for small shifts in λ. If the noise level is lifted, then more mixed series are generated. The region of ambiguity in the phase space, noted in the experimental results cited at the beginning of this paper, is wider if the noise level is raised, but there is a region of ambiguity between the almost single-state outputs associated with very low or very high λ (for given $\{a_k\}$ values).

All these results are for setting the interstimulus input interval long and for fixed a_k values, so they only correspond to a tiny region of the phase portrait where frequency and presentation rate are the axes used. As intrinsic noise is not a manipulated variable in the streaming studies cited, it is a new parameter of the system which could be relevant to the examination of individual differences in behaviour. Its presence is contraindicative for pooling data across subjects.

If the noise level is reduced virtually to zero (ϵ = .01) the the knife-edge effect is sharpened; the series before and after crossing $\lambda = 1.37$ in the example used are stuck in one state '1' for $\lambda < 1.35$ and stuck in '2' for $\lambda > 1.35$. The outputs are shown in Table 7.8.

Table 7.8: Sequences of Dominant Dimensions

90 inputs in series, 4Γ **Case 2** with Cascading

The shift from $\lambda = 1.35$ to $\lambda = 1.45$ with minimal noise.

$\lambda = 1.35, \epsilon = 0.01$

0 1 1 1 1 1 1 1 1 1 1 1 1 1 1 1 1 1 1 1 1 1 1 1 1 1 1 1 1 1
1 1 1 1 1 1 1 1 1 1 1 1 1 1 1 1 1 1 1 1 1 1 1 1 1 1 1 1 1 1
1 1 1 1 1 1 1 1 1 1 1 1 1 1 1 1 1 1 1 1 1 1 1 1 1 1 1 1 1 1

$\lambda = 1.40, \epsilon = 0.01$

0 1 2 2 2 2 1 2 2 2 2 1 2 2 2 1 2 2 2 2 1 2 2 2 1 2 2 2 2 1
2 2 2 1 2 2 2 2 1 2 2 2 1 2 2 2 2 1 2 2 2 1 2 2 2 2 1 2 2 2
1 2 2 2 2 1 2 2 2 2 1 2 2 2 1 2 2 2 1 2 2 2 2 1 2 2 2 2 1 2

$\lambda = 1.45, \epsilon = 0.01$

0 1 2 2 2 2 2 2 2 2 2 2 2 2 2 2 2 2 2 2 2 2 2 2 2 2 2 2 2 2
2 2 2 2 2 2 2 2 2 2 2 2 2 2 2 2 2 2 2 2 2 2 2 2 2 2 2 2 2 2
2 2 2 2 2 2 2 2 2 2 2 2 2 2 2 2 2 2 2 2 2 2 2 2 2 2 2 2 2 2

Table 7.9 gives some autocorrelations under two conditions, one on each side of the critical λ value used in Table 7.8. Programs in Fortran 77 by Chaghaghi (1985) were used.

The interest lies here is the absence of autocorrelation in the total output, but a marked local negative autocorrelation in the level of the maximum unidimensional output, which corresponds to what is called a *contrast* effect in some psychophysical studies of sequential judgments. This negative autocorrelation is damped in the series of differences between the total and the maximum component. There is little difference in this overall pattern, whether the process is on one side or the other of the knife-edge condition. The important feature to note is that the autocorrelation arises within the 4Γ process, it is not in the inputs. Because it is a predicted observable property of the outputs it constitutes a testable property of the model.

Table 7.9: Computed autoregressions

From zero lag to lag 9

$\lambda = 1.35, \epsilon = 1.10$

Y_{TOT} autoregression				
1.00000	-0.07223	-0.01101	0.01472	0.08126
0.02459	-0.03061	-0.03146	0.00635	-0.03760
$Ymax$(Re) autoregression				
1.00000	-0.22619	-0.03517	0.00670	0.02639
0.00528	-0.00531	-0.00291	-0.00802	-0.01924
δY_{TOT} autoregression				
1.00000	-0.12521	-0.02037	0.06807	0.07395
-0.14717	-0.06698	0.03398	-0.00912	-0.02067
$\lambda = 1.45, \epsilon = 1.10$				
$YTOT$ autoregression				
1.00000	-0.10465	0.00026	0.01161	0.01182
0.01770	-0.03863	0.02623	-0.01246	-0.04449
$Ymax$(Re) autoregression				
1.00000	-0.25475	0.00028	-0.00841	0.00970
0.00061	0.00628	0.01338	-0.01343	-0.01309
δY_{TOT} autoregression				
1.00000	-0.11898	0.01596	-0.01849	-0.04062
0.04166	-0.08308	-0.06469	-0.05808	0.08359

Note: The values of 1.00000 are merely in as a computational check.

Simultaneous Brightness Induction

The final case study presented here is one which, on the face of it, is a poor example to pick because it involves transductions specific to vision, and mechanisms whose physiology is not necessarily reflected in any general psychophysical model. Using this example is a way of seeing where generalities fail, and must fail, but trying to get some insights by seeing exactly how they fail. At the same time it is explicitly about a sort of field theory.

In brightness induction experiments, there are three visual fields, called Induction, Comparison, and Test. These can be arranged in various geometrical configurations, and in examples given by Heinemann (1972) they are all asymmetrical. The induction field may stand alone to one side, it may surround the test field only, or it may surround both the comparison and test fields. Fields may have a common boundary or be separated by an unilluminated region. Despite all this empirical variation the experiments utilising such configurations are directed towards a common objective, namely to investigate the effects of induction luminance upon the relative brightness of the other two fields.

One would have hoped, for simplicity, that Heinemann (1972) would have also reported work using a configuration such as a circle with three 120^0 segments which is in a sense balanced with respect to the weighting given to each of the three fields, and with respect to their pairwise adjacencies. However, because of the a *priori* status given differentially to each of the three fields experimenters have not treated the problem as the simpler one of three simultaneous luminance inputs interacting, where we ask for matching or relative judgments of brightness, and for unequal settings elicit responses to each component field in turn.

More recent studies have been concerned with more complicated displays, typically involving stripes, patches, and special edge effects. It appears that by so doing more than one sort of mechanism is induced to operate; Moulden and Kingdom (1989) conclude that one mechanism mostly operative at corner intersections, and another spatially extended, are both involved in the perception of gratings. In this treatment we are only concerned with extended mechanisms, though the network model of chapters 5 and 6 obviously refers to corner intersections. Extended mechanisms are there only covered by repeated diffusion through cascades.

The starting point of this theoretical analysis is a graph presented by Heinemann (1955, 1972) in which are plotted a family of curves, all of a similar shape, on the luminance axes $logL_i, logL_t$, of equal brightness, i.e. $\psi L_c = \psi L_t$ for a series of increasing $logL_t$ values in millilamberts. This is recreated in Figure 7.17, in part. The important feature is the sharp drop-off of the matching curves; Heinemann (1972) observes "Near the point where $L_i = L_t$ small increases in L_i make necessary very large decreases in matching luminance, L_c, and when L_i exceeds L_t by only a few hundredths of a log unit the brightness of the test field is reduced so much that a match is no longer possible". he also notes that "The nearly linear portions of the curves have slopes near 1.0. This indicates that the brightness of the test field is almost constant when the ratio L_t/L_i is constant."

This means the plots considered are only possible within a limited range of luminance parameters, and their form may be indeterminate when their rate of change becomes almost infinite. Heinemann uses a log range for L_1 and L_c from 10^{-2} to 10^2 which implies an extreme scale compression in the region of the upper L_i parts of the curves.

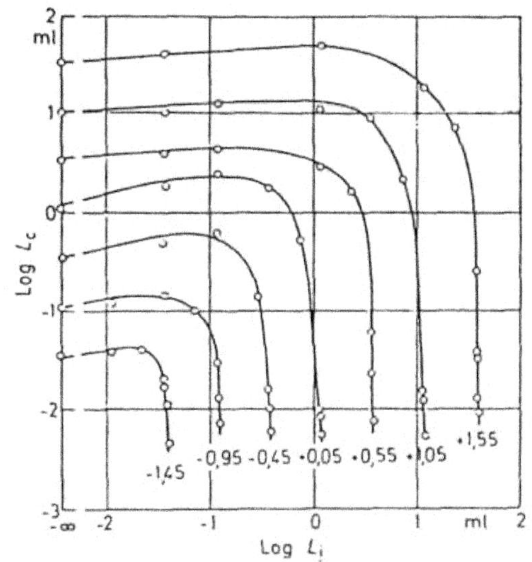

Figure 7.17 adapted from Heinemann (1972)
$\psi L_c = \psi L_t$ **as a function of** L_i, L_t

Theory

Some notation is now introduced so that empirical work can be matched to the nonlinear psychophysical equations, using the 3Γ **Case 2** model.

The physical variables are:

$$L_i = \textit{induction field luminance},$$
$$L_c = \textit{comparison field luminance},$$
$$\textit{and } L_t = \textit{test field luminance}$$

In plotting empirical functional relationships $logL$ values are conveniently used, but we will initially take the L values as defining the implicit U inputs to 3Γ, so the a_i, a_c, a_t values are monotone increasing transformations of L when we come to simulate the topography of Figure 7.17.

As in the algebra of 3Γ the interpretation of a_i, a_c, a_t is immaterial, the relative values and their cross-coupling alone carry the burden of simulation of the relationships in Figure 7.17, we will drop the c, i, t suffices and simply work with a_1, a_2, a_3.

In matching experiments, the problem of the metric of the brightness response variables is circumvented, because equality of perceived brightness is defined in terms of a matching of potentially unequal $logL$ stimulus values. We wish to reintroduce the outputs Y_1, Y_2, Y_3 of 3Γ which would correspond to $\psi L_i, \psi L_c, \psi L_t$ respectively, that is the perceived brightnesses of the three fields. As

$$\psi L_i = induction\ field\ brightness,$$
$$\psi L_c = comparison\ field\ brightness,$$
$$and\ \psi L_t = test\ field\ brightness$$

the correspondences are given in Table 7.10:

Table 7.10: Mapping of Parameters between Data and Theory.

$$Data \Leftrightarrow Theory$$
$$Luminances\ \ 3\Gamma\ inputs$$
$$L_i \Leftrightarrow a_1,$$
$$L_c \Leftrightarrow a_2,$$
$$L_t \Leftrightarrow a_3,$$
$$and\ Brightnesses\ and\ outputs$$
$$\psi L_i \Leftrightarrow Y_1(Re),$$
$$\psi L_c \Leftrightarrow Y_2(Re),$$
$$\psi L_t \Leftrightarrow Y_3(Re),$$
$$so\ L \mapsto \psi \Leftrightarrow a \mapsto\ Y(Re),$$

There are no variables in the luminance measures corresponding to the λ parameters in 3Γ, but the idea of mutual (and possibly asymmetrical) interference between luminance fields is taken up by λ.

Theoretically, it is intended that in a display L_i, L_c, L_t the induction field is presumably resistant to 'masking' or some sort of interference from the test and comparison fields, but the latter two are susceptible to interference from L_i and also mutually interact. We relax these suppositions, and simply assume that the capacity of L_c and L_t to interfere is *symmetrical*, even though it is usual only to elicit a response to L_c. It is possible to do these experiments with two inputs to one eye, and the induction to the other, which implies that the interaction of inputs is central and not retinal. As NPD is postulated to be concerned with central and not peripheral processes it is reasonable to explore the modelling of brightness induction from this standpoint. Turning this into a 3Γ **Case 2** model, it is implied that we set

$$\lambda_1 > \lambda_2,\ \lambda_1 > \lambda_3,\ \lambda_3 \sim \lambda_2 \qquad [7.14]$$

so the 3Γ model is now asymmetrical and in full has six parameters

$$\lambda_1, \lambda_2 \sim \lambda_3 = \lambda, a_1, a_2, a_3, \eta \qquad [7.15]$$

If we were directly to try and simulate Figure 7.17 curves, then the requirement would be to find loci of the relation

$$a_2|(Y_2 = Y_3) = f(a_1, a_3) \qquad [7.16]$$

For the purposes of simulation, set $\lambda_3 = \lambda + \delta\lambda$, and treat $\delta\lambda$ as a variable. The other variable is $dY_{2,3} = |Y_2 - Y_3|$, which is zero for a matching experiment in the sense of Heinemann (1955).

Simulation

We now move to create a contour plot of $dY_{2,3}$ over the whole a_1, a_2 field, for each a_3 value. Initially, because of the extreme range of luminance values used in the Heinemann (1955) data, it appears expedient to explore the whole stable a range within the Γ recursion, that is from just above $a = 2$ to just below the chaotic and explosive region around $a = 5$. That is, we treat matching as a special zero local region of the dY output, and see what shape it takes within the total a_1, a_2 field. This gives a complete 2-d areal plot for each of Heinemann's lines.

If the 3Γ model, for some parameter settings, functions analogously to the brightness induction experiment, then as a formal hypothesis

If 3Γ, $\forall\, a_3\ \exists\ \lambda, \delta\lambda \mapsto$ *a trough in* $dY_{2,3}$ *convex on the* a_1 *axis*,

or in other words the brightness induction data are the partial realization of a 3Γ system.

This raises the question, why do the matching experiment at all, instead of the $dY_{2,e}$ plot on a_1, a_3? The matching is a variant of the method of limits, which can have multiple solutions (Gregson, 1992b). If the operation of matching does not induce some special psychophysical mechanisms, what has been called a 'bias' by Poulton (1989), then the locus of the trough is preserved. But if asking for mapping zero differences, as opposed to asking for the size of a difference, induces some secondary mechanism, then the trough is displaced in the field. This problem is reminiscent of peak shift in operant conditioning, or optimum shift in experimental aesthetics. Another representation is available, if entropy plots are used, when the maximum entropy regions should be on the shoulders of the trough.

A series of contour plots of the $dY_{2,3}(\mathrm{Re})$ surface are now examined; they are shown in Figures 7.18 through 7.27.

Table 7.11: Parameters Settings for the 8 Simulations.

Contour Maps of the $dY_{2,3}(\mathrm{Re})$ Surface.

Figure #	a_3	λ	$\delta\lambda$	limits on a
7.18	3.2	1.3	0.4	unbounded
7.19	3.0	1.3	0.4	bounded
7.20	3.2	1.3	0.4	bounded
7.21	3.2	1.3	0.6	bounded
7.22	3.7	1.3	0.4	bounded
7.23	4.2	1.3	0.4	bounded
7.24	3.2	0.9	-.1	bounded
7.25	3.2	0.8	-.2	bounded

It is apparent from comparisons within the series of simulations presented that only in a narrow region of the 3Γ parameter space are generated contour maps with rapid roll-off for high a_1.

The sharp roll-off is only produced if the a values are all bounded, arbitrarily here at 5.5, in order to avoid the generation of explosions in at least one Y output at some locations in the 25×25 matrix. This bounding only applies to the a values when they define the gain and not to the e values which are defined in 3Γ as $\lambda \cdot (max(a))^{-1}$. This means that the gain on dimension 1 is bounded, but the e_2, e_3 values continue (for fixed λ) to drop off as $U_1 \rightarrow a_1$ input increases. This in turn modifies the Y_2, Y_3 values but not the Y_1 values, so that $dY_{2,3}$ exhibits the roll-off.

Interpreting this in terms of brightness, the central representation of any field luminance hits a saturation level, or (which is the same thing in this context) the transmission of the representation of luminance between peripheral and central levels has an internal upper bound. We illustrate this by comparison of Figures 7.18 and 7.19; the remaining argument rests only on examination of Figures 7.19 through 7.25.

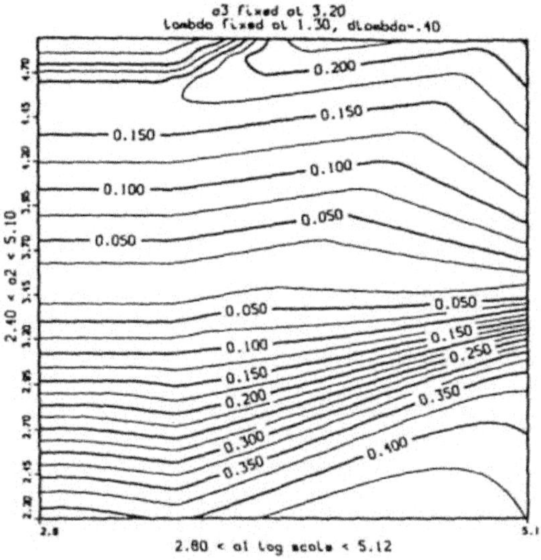

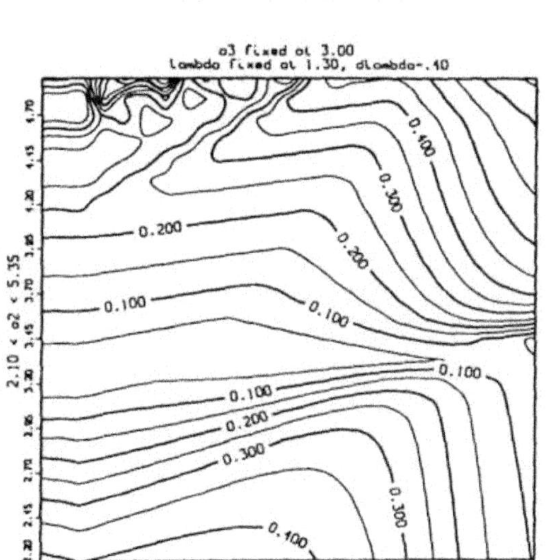

Figures 7.18 and 7.19 simulations in 3Γ

Contours of $dY_{2,3}(\mathrm{Re})$ with various a_3 and various λ and $\delta\lambda$

It appears that bounding the a values, which also implies a constraint on the e values is a necessary condition for the roll-off to occur, but it is not a sufficient condition. Figures 7.20 and 7.21 show the results of increasing $\delta\lambda$, which represents the asymmetry of mutual interference of the three channels. The effects of $\delta\lambda$ are second-order, but increasing $\delta\lambda$ does very slightly sharpen the roll-off effect.

Consider Figure 7.21; the roll-off occurs both above and below the middle levels of a_2, but not at the middle range of a_2. That is, the region where the $dY_{2,3}(\mathrm{Re})$ surface goes to zero (the matching condition of Heinemann) is the only region where the contours are almost horizontal up to the right-hand margin. The simulation reproduces roll-off everywhere (in terms of $dY_{2,3}$) except in the region of attempted precise matching that the visual data exhibit. At the right margin the region of near-zero $dY_{2,3}(\mathrm{Re})$ is very narrow; the surface gradients are steep.

One possible reason for the brightness data not corresponding to theory in the exact matching region is that there is hysteresis in the matching operation itself, so that the observer rarely stops on an exact matching, but always a bit to one side or the other; this is what happens in the psychophysical Method of Limits and the p.s.e. is obtained by averaging responses which asymmetrically under- or over-shoot the correct match. This would constitute a second mechanism between peripheral and central levels which could account for the curious disparity between simulations and and raw (or better, partly presented) data.

It should be noted that the top boundary of some of the graphs goes into turbulence, to varying degrees; it is therefore predictable that a chaotic region, for matching, would arise for very low a_1 and high a_2, for most a_3. The Heinemann (1972) data are very sparse in this region so that there exist no empirical checks on the prediction, which in any case would show itself only as very high response variance (i.e. low reliability) in that subregion, if it existed.

We now turn to Figures 7.22 and 7.23, which show results of pushing up a_3 for fixed λ, $\delta\lambda$. The roll-off effect vanishes. The region of matching goes up the graph so that matching occurs with greater a_2. This would mean that at high test field brightness the whole dynamics change, a_1 (induction) ceases to have any differential effect and a_2 (comparison) values are less critical. Whether this is testable without actually damaging the retinas of the observer is not clear. The algebra can make predictions for system parameter values that are not physiologically realizable.

Figures 7.24 and 7.25 are the consequence of setting $\delta\lambda$ negative, that is the relative values of mutual cross-coupling between dimensions are reversed.

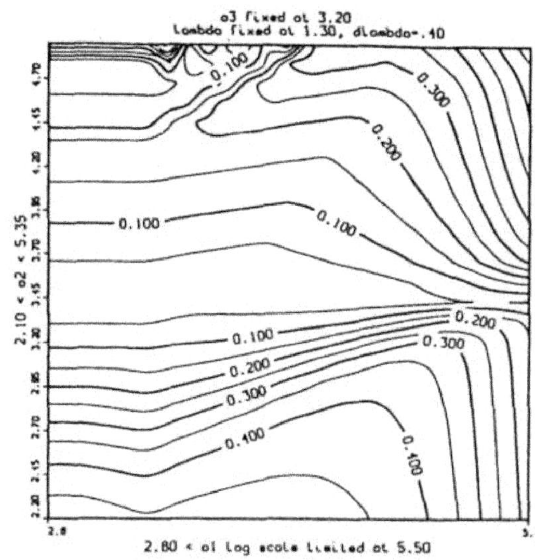

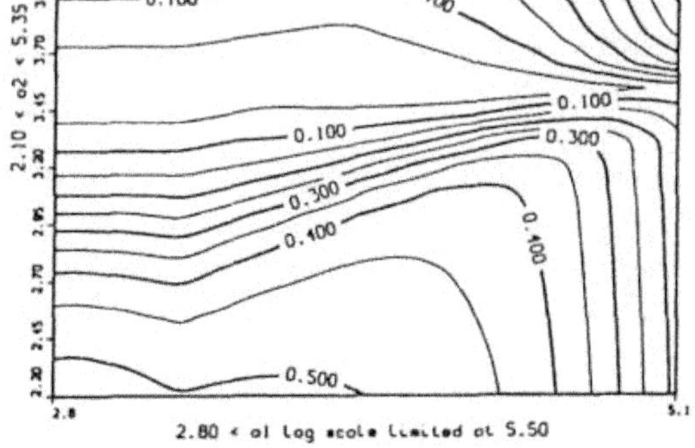

Figures 7.20 and 7.21 simulations in 3Γ
Contours of $dY_{2,3}(\mathrm{Re})$ with various a_3 and various λ and $\delta\lambda$

In order to avoid chaos and explosion it is necessary to set λ values lower than for the previous figures. Some roll-off occurs again, in a limited region of a_1, a_2; the curiosity now lies in the lower right hand region of the graph (high a_1, low a_2) where a large region of insensitive matching ($\delta Y_{2,3} \rightarrow 0$) is created. In these hypothetical cases there would be no exact matching of $Y_2(\mathrm{Re}), Y_3(\mathrm{Re})$ for high a_1; anything would do within limits. Again, changes in $\delta\lambda$ are not critical within a range of values that avoid explosion.

In all the simulations, η (the terminal step of the Γ trajectory) has been fixed at 10; following the precedents of most earlier $n\Gamma$ studies for computational convenience. However, if the 3Γ process is pushed into limit cycling on at least one of Y_2, Y_3, which will correspond to the upper left corner of a graph (as for example in Figure 7.24) then two surfaces, for η, $\eta + 1$, would exist and envelope the $dY_{2,3}$ function. The whole map

$$(a_1, a - 2, a_3) \longmapsto (Y_1, Y_2, Y_3)$$

is expected to be heteroscedastic in terms of $\nabla dY_{2,3}$.

We conclude that some of the features of simultaneous brightness induction are not necessarily the consequences of peripheral visual processes at all, but could be common to any central processing in which (a) we have a 3Γ network, (b) there are strong bounds on the range of transmitted inputs to the central levels, and (c) the method of data collection exhibits hysteresis. It is obviously easier to perform such experiments in the relatively simple and manipulable case of vision and brightness, and finding tractable analogues in other sensory modalities would demand considerable experimental ingenuity. It is suspected that an analogue in tactile multistimulus experiments might be created.

Discussion

It would appear that there does exist a region in the 3Γ parameter space that generates observable relations between inputs and outputs (with arbitrary scaling) comparable to the brightness induction phenomena, with the extreme scale compression in a_1 (and to a much lesser extent in a_2). However, it should be noted that variations in $\delta\lambda$ around zero may not be critical; Using a negative $\delta\lambda$ would imply a situation where the test and comparison fields modify the induction field brightness more than their brightnesses are modified by the induction luminance. As the literature does not usually address this question (indeed from the standpoint of brightness experimental designs it may be thought "the wrong thing to ask") we have no comparable data.

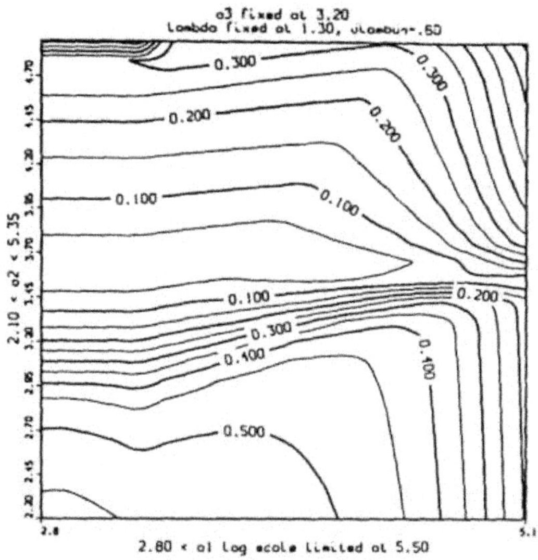

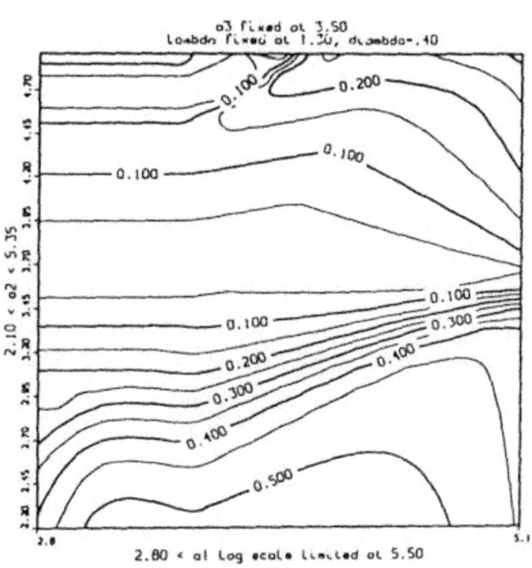

Figures 7.22 and 7.23 simulations in 3Γ
Contours of $dY_{2,3}(\mathrm{Re})$ with various a_3 and various λ and $\delta\lambda$

The phenomena we are aiming to reproduce are indeterminately a mixture of peripheral and central processes, and possibly also a consequence of the demand characteristics of the matching tasks created by experimenters to look at brightness induction. Any peripheral process is in effect prefiltering on what happens centrally; in the sense in which Γ transformations are used, the periphery constrains the ranges of the a parameters, and the demand characteristics locate the output in the parameter field. The scale compression effect might be peripheral and specific to brightness vision, but the general nonlinear mapping located centrally as a common multiple input processing. If this idea is tenable, then the $n\Gamma$ theory can be used as one more tool to disentangle central and peripheral psychophysical transformations within a given sensory channel.

Brightness induction is now considered (Foley and McCourt, 1985) to be one of a group of visual effects in which the mapping of luminance with brightness is not simple, that is, it is not a linear transformation. Recent work has shifted attention to more complicated situations than the paradigms reviewed by Heinemann (1972) and "brightness induction may be another special case with spatial frequency equal to zero"(Foley and McCourt, 1985, p 1220).

In other words, brightness induction using three areas, each homogeneous in luminance, is the simplest case where frequency resolutions do not further complicate the situation, nor need explicit representation within a psychophysical model. It is for this reason that a general modality-neutral psychophysics might predict some relations between luminance and brightness variables in the simplest cases, but be useless with much more intricate stimulus inputs.

It does not follow that we can safely extend the argument unconditionally to any other sensory modalities, without consideration of what occurs when multiple inputs are involved. For example, in auditory psychophysics (Zwicker and Fastl, 1990, chapter 14) copresentation of two frequency inputs creates quadratic or cubic combination tones and the system is nonlinear in more ways than directly derivable from a 2Γ or 3Γ model; there frequency and amplitude are interlocked.

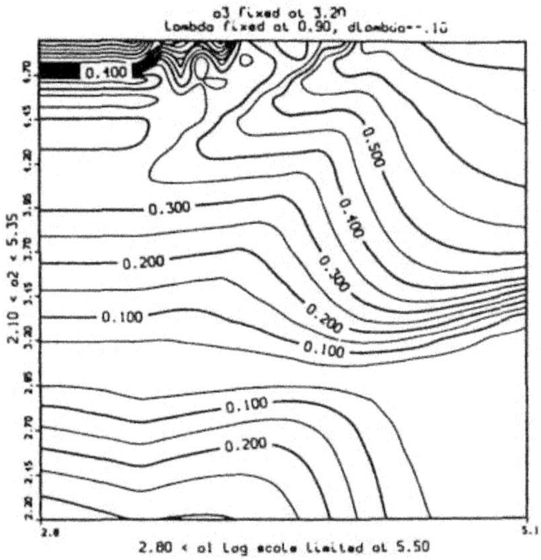

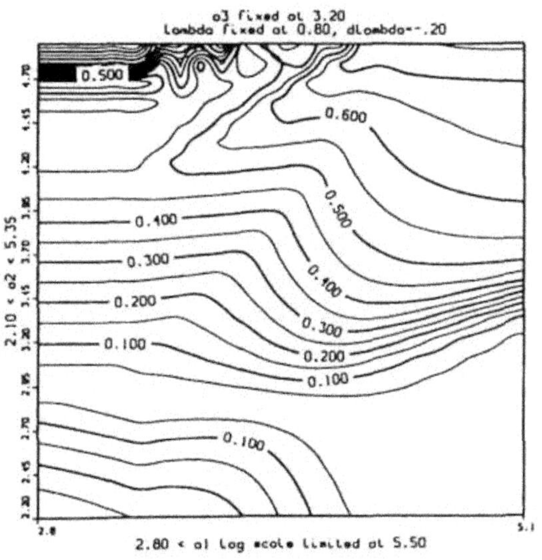

Figures 7.24 and 7.25 simulations in 3Γ
Contours of $dY_{2,3}(\mathrm{Re})$ with various a_3 and various λ and $\delta\lambda$

8 Postscript: Psychophysics or Events in the Real Brain ?

"...the whole system does not look like a one-shot, static, response mechanism. It is likely to operate by many transient dynamic interactions, conducted at a fairly fast rate."

F. Crick, *The Astonishing Hypothesis*, 1994, p. 159.

A resurgence of interest in the relationships between conscious awareness, and events in the brains of the higher apes and man, has produced theories by physicists, neurophysiologists, and psychologists (Bock and Marsh, 1993). Most of these are frankly speculative, and the worst built on a complete ignorance of relevant evidence from psychophysiology. A valuable exception to this cheerfully undisciplined theorising is provided by Crick, and the quotation at this chapter's head provides a bridge between the neurophysiology he reviews, with meticulous attention to collateral results in the psychology of visual perception, and the nonlinear psychophysics sketched out in this and the two preceding monographs. For the purposes of this postscript, results in all three are drawn on.

It is salutary to remind ourselves that at the core of psychology is still the body-mind problem in some guise, and that mapping from physically-defined variables to the intensity or patterning of sensory experience is a terse shorthand (and often little more) for the path between the outside world and the mind, via undisclosed pathways in the brain. In practice, it is only the invariances in such a mapping that can be captured in algebra even if it is, as here, the algebra of nonlinear dynamics floating into and out of chaos.

Traditional psychophysical equations, such as those invented by Weber, Fechner, and Plateau, are one-shot and static; they tell us about input-output regression equations that can be plotted when residual error is partialled out as Gaussian noise, and when sufficient monotone transformations on physical and on psychological numbers, or associated probabilities, lead to some tractability in the form of restricted linearity over tight stimulus ranges. In no sense are they process models which could match the characteristics of brain dynamics identified by Crick. There are many historical reasons for this situation, even though Fechner was trying to create an "inneren Psychophysik" which had its putative roots in physiology (Murray, 1993). In contradistinction, the family of models used here built around one and the same Γ core, by imposing boundary conditions or ancillary assumptions on its evolution, is not one-shot nor static, but dynamic and very transient. Whether or not these negative and positive attributes, created completely independently of Crick's approach, match is a sufficient sense some of the mass-action dynamics of the brain is still a very open question. What can be said is that at a metatheoretical level there are compatibilities between the description of some brain dynamics that go along with conscious experience, and the intricate shifts and instabilities deliberately engendered some of the time in the Γ family of models.

A contrast between the static structure of classical psychophysical equations and the input-output mappings of a Γ process is illustrated by considering what affects the rate of change in the psychometric function, in the two approaches. Using

$$\psi = a \cdot \phi^b + \epsilon, \quad \epsilon \sim N(0, \sigma_\epsilon) \qquad [8.1]$$

which is the so-called psychophysical power function, the slope $d\psi/d\phi = f(b)$ whereas for Γ there is no equivalent invariant stimulus-response form, but

$$d^2Y(\mathrm{Re})/da^2 = f(e, \eta) \qquad [8.2]$$

and for low e, $d^2Y(\mathrm{Re})/da^2$ *is* $+ive$ but for high e, $d^2Y(\mathrm{Re})/da^2$ *is* $-ive$. The characteristic ogival form often found is associated with intermediate values of e. The point is that the Γ family of processes are not written as a view from outside the black box, as is [8.1], but rather as processes within the box, from which as a secondary consideration a number of predictions about stimulus-response equations having no reference to duration of processing, nor to sequential effects, can be derived if they were wanted to compare with real data. It is the lability of nonlinear processing underlying [8.2] that brings it one step nearer to the global neurobiological dynamics of the brain.

In this volume the core of the trajectories which are cascaded, networked, or cascaded and networked, has been for simplicity just one of the family called Γ$V7$ in the previous two monographs. To illustrate what can happen this is sufficient, but to model some sorts of pathological experience associated with brain lesions some other models, such as $\Gamma V1$ or $V4$ might be better. This point is mentioned here because evidence from patients with rare types of localised lesions plays a critical role in discussions about the brain activity that is necessary and sufficient for consciousness to be observed as first-person data, or for reports of consciousness as third-person data.

Distinctions have to be drawn from the start, between the properties of the psychophysical models here, and other dynamics studied in theoretical physics, such as cellular automata or spin-glass networks, which mathematically may appear to be related. Some comments on this point were made in Chapter 1. Other distinctions have to be drawn from parallel distributed processing networks, though in a loose sense analogies are discernable. As Γ trajectories have potentially intricate dynamics their identification can, in very long sequences, be characterised by the usual indices, such as Renyi D_2 (the correlation dimension of fractals) or the Lyapunov exponents. To see what sort of analytical mathematics are implicit in the Γ dynamics if those dynamics are persistent in time and not transient, these indices and Julia sets with fractal boundaries can be generated. Examples are given in the 1992 text, but to show further the lability of the dynamics Table 8.1 gives some approximate computed estimates, for runs 1000 iterations long, in the region where the bifurcation sequences lead in to aperiodicity just short of explosion.

Table 8.1: Attractor parameters for Gamma

For $Y(\text{Re})$, $\eta = 1000$, Renyi D_2 (correlation dimension), and the Largest Lyapunov Exponent, from Wolf *et al* (1985) algorithm.

a	e	D_2	max(Lyap)
4.2	.460	1.37	-9.02
4.2	.471	1.27	$\simeq 0$
4.2	.473	1.37	+2.75
4.7	.320	1.88	+6.29
4.7	.340	1.44	-10.88
4.7	.360	1.12	-2.19
4.7	.380	1.48	+3.75
4.7	.388	1.03	+4.32
4.7	.400	1.06	+2.78
4.7	.420	1.20	+4.84
5.2	.300	1.25	-5.00

Table 8.1 is a reminder that the shape of the attractor varies with the system parameters in such an intricate way that, observed from outside the black

box, it looks almost unstable and uncontrollable, even though it is deterministic. Only the indices of $Y(\text{Re})$ have been shown; the corresponding values for the $Y(\text{Im})$ component of the dynamics can be quite different. For large e values the $Y(\text{Im})$ trajectory is effectively random noise, the D_2 value is large (> 200). For some a, e pairs the two parts of Y are in phase and period two, for others $Y(\text{Re})$ can be periodic and $Y(\text{Im})$ aperiodic, and the rate at which the components settle to periodicity when they do so, is not the same even if they end up in phase. Effectively $Y(\text{Im})$ looks like background noise, because its D_2 is like that of an almost infinite-dimension attractor, but it does not always resemble Gaussian noise, and may die out by Y absorbing onto the real line. Generally one can say that $Y(\text{Re})$ has a $1 < D_2 < 2$, which is unsurprising[1]. If one wished to compare Γ dynamics with mathematical physics this is one way to go, but this is most certainly not the point. The long-term mathematics which are revealed, sometimes, with computer simulations running to millions of iterations in order to produce beautiful images of attractors, miss the point completely *within the present scenario*, they are about something else. In all the applications of nonlinear psychophysics which have been explored, the trajectories are constrained to run for a very brief burst and then terminate, by jumping out, by restabilisation, by cascading to another transient, or by being overwritten by new dynamics at a different level, or even by being coarse-grained into Markov chains. Examples of all these are in the referenced publications. In the psychophysical world, where subjects get their conscious experiences — if they get them at all — fairly quickly after being stimulated by the onset of events impinging on receptors, it is the dynamics during destabilisation and the first few recursions asymptotic to stability which have been created and explored in the models. In a few extreme cases, therefore, associated with high ae products in the Γ recursion, trajectories have been used up to $\eta = 10$ or to $\eta = 20$ which, if carried further, would have exploded. Restabilisation has to come in quickly. A range of phase space portraits have been covered by the dynamics, from rather uninteresting convergence onto point attractors, to interactions between channels in which one can be chaotic and the other strictly periodic. In networked explorations of fields the latter cases are particularly rich and suggestive. Certainly what has been created and run is a family of processes, taking time to evolve between stimuli and potential responses, which are, in

[1] Note that the D_2 of the human EEG typically lies between about 6 and about 10, which is much higher. Γ dynamics are not equivalent to long run fluctuations in mass-action of the brain measured on the scalp.

Craik's words, transient dynamical interactions conducted at a fairly fast rate.

But are they in any identifiable sense in tight correspondence with underlying brain events ? Certainly the psychophysics here is not about the performance of single neurons, or even simple local collectives of networked neurons. If parallels are sought between some brain psychophysiology and $n\Gamma$ dynamics, then a single Γ trajectory is at the least a model of the performance, but not the structure, of two coupled recursive pathways, interlocked and with quite different rates of convergence to stability. If $n\Gamma$ and $(n \times n)\Gamma$ are used, then the number of pathways needed to compute the algebra is at least $2n$ or $2n^2$ respectively for each cascade step. Sensory processes have pathways simultaneously activated in the thalamus and the neocortex, and various examples of this have been traced out physiologically[2]. Some recursive pathways have special functions; oddities of temporal and spatial perception arise when these pathways are lesioned. In vision, several pathways are typically involved, and recursive activity between cortical levels is ubiquitous if transient. The question, then, is are the $n\Gamma$ dynamics isomorphic, at a system theory level, with a subset of brain processes which mediate conscious sensation ? The NPD approach is, by definition, a sort of cybernetics, but obviously like all psychophysics an incomplete one.

In this study use has been made of the cascading process, in which the dynamics resemble the pushing of activation up through a hierarchy of levels, using the repeated $a \mapsto Y \mapsto a \mapsto Y$..... transformations. Constants are used to keep this within bounds, but the key idea is simply the higher-level recursion. This too has some correspondence with brain activity, Crick notes that the neocortex is "characterised by its ability to evolve additional layers in its hierarchy of processing".

Comparisons with Hippocampal Functions

The dynamics of $n\Gamma$ cascades involve three fundamental properties, internal Γ recursions (in j steps), external recursions (in η steps between k cascades in series), and a periodicity (in Y) intrinsic to the dynamics of the internal

---

[2] Wall, P. D. (discussion in Bock and Marsh, 1993, p. 295): "A sensory function may reach a single conclusion such as "my foot hurts me" but it has done so by a sequence of processes distributed in space and time. All the experimental evidence now points to the process (and for that matter, the conclusion) being an interactive pattern in time and in many structures, both cortical and deep."

recursion in some parts of the system phase space. There can also be forcing functions from external inputs onto a, so that, for example, $a_k = f(cos(k))$, but it has nowhere been necessary to use these or to consider their effects in the applications so far studied. From the perspective of an external observer the input-output mappings can lead to stable or to unstable activation patterns within the process trajectories, as has been explained by examples in some detail.

The question however is whether such dynamics, even in their simplest form, undifferentiated between sensory modalities, have any parallels in the sensory neurophysiology which underlies psychophysics. Nonlinear psychophysics is mathematical modelling of the regularities in sensory or perceptual data, it is not psychophysiology *per se* but it is implausible if its postulated dynamics have little correspondence to what is known the brain can do and does do when it receives sensory inputs. One is not obliged to adopt a strong reductionist stance in psychophysics, but the theorist has always to remember that psychology is a bridge between levels of analysis, behavioural and neurophysiological. The relevant data are not always known for us to pursue this question in satisfactory detail, but the most promising results arise from modern knowledge about the functioning of the hippocampus and connected adjacent structures in the entorhinal cortex which are the source of major inputs to the hippocampus.

As we are only concerned with sensory mappings and not with learning, and as abstract psychophysics is written in terms of an undesignated modality or channels, the mappings from the hypothalamus-amygdala circuits onto neocortical areas with specific sensory function are not considered, though in learning to name and verbally or otherwise behaviourally to discriminate sensory qualities, it would be expected that they are integral to the processes in establishing what has come to be called long-term potentiation (Macphail, 1993).

We must consider both intrinsic and extrinsic circuits involving the hippocampal sensory functions. The anatomical foundations for closed loops exist in the hippocampus, which together with the role of these structures in theta ($\simeq 5 - 7\ Hz$) rhythms in the electroencephalogram, provide an analogue to the Γ dynamics. The Papez circuit involves afferents to the hippocampus from the septal area, and these are necessary for theta activity (Macphail, 1993, p. 262). It is, particularly, the entorhinal cortex rceiving inputs from the hippocampus which is of interest where analogies to $n\Gamma$ might be sought. Four major multimodality projection areas of the entorhinal cortex exist; this cortex does not receive direct projection from unimodal association cortices, and this pattern is taken to suggest that the hippocampus is not concerned with information arising at any specific modality, but rather with abstract and analytical information from any and all sensory modalities. Olfaction might be an exception because of its direct innervation to the hippocampus.

The hippocampus is not purely a feedforward information processor, however, it has outputs both to the multimodal areas and to unimodal parasensory association cortices. Such a substrate can support a dynamical system in which the hippocampus not only monitors input already processed, but modulates its own intake of information. In a very simplistic and approximate sense, this is what the e and λ parameters of $n\Gamma$ do, and if they are modified as a function of feedback within a cascade, and not treated at least locally as constants, then the dynamical complexity of the theory moves into closer correspondence with neuropsychological reality.

It is unsurprising that neural network models of the hippocampus, of the PDP structural type mentioned in Chapter 1, have been constructed to represent learning, involving both inhibited feedforward and feedback (Rolls, 1989). Of particular interest is a model by McNaughton and Morris (1987) because it can be shown that an network analogue of the hippocampus can perform division operations on simultaneous inputs (as used in the definition of similarity in Chapter 7) and also behaves like a type of correlation matrix. It appears that such a network will support the autocorrelation of sequential but not parallel inputs. The argument against taking such models as too literal a structure for psychophysics rests on the problems of saturation which arise in a network with as few as six channels; the information handling capacity, for both quantitative and qualitative information about stimuli, does not closely parallel what the human observer can do. There are two other aspects of hippocampal function that might relate to the use we have made of $n\Gamma$ in the perception of both time and of spatial patterns. From an NPD approach it is both fortunate and curious that many cells in the hippocampus encode a diversity of functions; some theorists have suggested that spatial maps exist not only in the visual cortex but also hippocampally, and an intact hippocampus appears to support the retention of at least spatial relationships at the phyletic level of rat. Learning to respond after a time delay is, in rats, a process which demands an intact hippocampus, and the internal clocks controlled by cholinergic input to pacemaker cells can apparently be suspended in their action briefly in some way.

It seems at least uncontentious to conclude that the neurophysiology, and indeed the cytoarchitectonics of sensation do interlock the various dynamics of single and multichannel psychophysics, of spatial networks, and of time-based behaviours that have also been linked in this study by the use of Γ, $n\Gamma$ and $(n \times n)\Gamma$ and cascades, within the same theoretical system. If this analogy is not fortuitous then it might go some way to explain two things, why the gross dynamics of Γ systems can mimic psychophysics, and why the same algebra of nonlinear dynamics can serve to illuminate a diversity of sensory physiology which extends into what were regarded as perceptual and even cognitive, and not simply sensory, as psychologists traditionally viewed them.

Visual Consciousness

Perhaps the greatest temptation, to look for parallels, is between the anatomy of the visual cortex and the $(n \times n)\Gamma$ cascades used in Chapters 5 and 6, but extreme caution must be used here. If it is conscious visual experience that is being sought, then it is salutary to observe that so far no single neuron region has activities that correspond to our picture of the world. As it has been shown that some of the patterns in Γ networks match some Gestalt phenomena, at least topologically, then functionally but not in microphysiological terms there might be some correspondences. The global mixed serial and parallel functioning of the brain over a wide volume of tissue, and elicited by a diversity of cognitive tasks, allows us to postulate such mixtures in the cascaded networks as being intrinsically feasible as wetware computations.

The comparisons between $(n \times n)\Gamma$ evolutions and some properties of visual neurophysiology and conscious awareness are perhaps surprising for the closeness of the resemblances which result. In reviewing some non-veridical effects in the simplest of geometrical illusions, it was shown that the network activity corresponding to the stimulus line figures could persist, fluctuating but relatively dominant over transient patterns in the surrounding ground. The stronger and more persistent distortions of the figure had some resemblance to the phenomenology of well-known Gestalt effects, the weak transients in the ground do not match reliable findings in the experimental psychology of conscious perception. To find what is stable and invariant in the $(n \times n)\Gamma$ cascaded networks two approaches were tried; stopping the cascade at an early iteration, or in effect integrating over the local system parameter space (λ, η) to smooth the ground effects which are more sensitive to small parameter fluctuations. They then average out and vanish, leaving only a corrupted figure. But the initial evolution of the lattice may have its counterparts in pre-awareness processing, for as Libet (1993, p. 131) observes, " the transition between psychological detection of a sensory signal without awareness and the detection with awareness can be controlled simply by the duration of the repetition of similar ascending activations of sensory cortex." This in fact holds for either sensory thalamus or sensory cortex.

These global dynamics are interesting because they can be generated using as the one core a Γ recursion which is the same as used in the one-dimensional sensory psychophysics developed for previous less complicated applications of the theory. This sort of parsimony does have metatheoretical implications, no extra basic mechanisms are invoked for each new class of sensory-perceptual phenomena, In evolutionary terms, the brain is not built with more than it needs to support a diversity of tasks leading, at the highest phyla, to consciousness in its fullest sense.

Internal Delay in the Gamma recursion

Throughout all the analyses in this and the preceding monographs, the cubic recursion of an internal state complex variable Y, within the time interval of one trial or one cascade step, in discrete time units j, has been written and generated as

$$\Gamma: \quad Y_{(j+1)} = -a \cdot (Y_j - 1)(Y_j - ie)(Y_j + ie) \qquad i = \sqrt{-1} \qquad [8.3]$$

However, this assumes that the recursion is over one and only one step, from j to $j+1$. This is a structural assumption that can be relaxed, and from evidence about variations in internal delays within neurological circuits, the consequences of its relaxation should be noted; they turn out to be dynamically complicated in the mathematics. Delay in sensory information processing does, of course, happen observably when brain function is impaired in alcoholism, dementia, or depression, to name only a few states of clinical importance.

Guckenheimer and Holmes (1983) looked at the consequences of rewriting the logistic equation in the real x to be, instead,

$$x_{j+1} = \mu x_j(1 - x_{j-1}) \qquad [8.4]$$

which is called (Thompson and Stewart, 1986, p. 159) the 'delayed logistic map'. Obviously the Γ recursion is more intricate, and introducing different delays calls for more detailed considerations, though the changes even in the simpler logistic equation in rewriting it in its delayed form are mathematically subtle. There are changes in what Block and Coppel (1986) called the stratification of the maps on the interval. Introducing some notation, d is a delay,

Delay Operator Realisations

0,0,0	$=\omega 1$	0,1,1	$=\omega 5$
0,0,1	$=\omega 2$	1,0,1	$=\omega 6$
0,1,0	$=\omega 3$	1,1,0	$=\omega 7$
1,0,0	$=\omega 4$	1,1,1	$=\omega 8$

$d = 1, 2, 3...$ only taking integer values, and ω is a delay vector operator with three elements, each of which is zero or one.

Some of these forms appear to be mathematically so unstable as to be useless for our purposes. Now rewriting Γ for two cases of interest as

$$^{d}\Gamma_{\omega 3}: \quad Y_{(j+1)} = -a \cdot (Y_j - 1)(Y_j - ie)(Y_{j-d} + ie) \qquad [8.5]$$

and

$$^{d}\Gamma_{\omega 5}: \quad Y_{(j+1)} = -a \cdot (Y_j - 1)(Y_{j-d} - ie)(Y_{j-d} + ie) \qquad [8.6]$$

and noting that

$$^{d}\Gamma_{\omega 1} =_{def} \Gamma \qquad [8.7]$$

and that $^{d}\Gamma_{\omega 8}$ is the nearest parallel to the delayed logistic form, we may examine the dynamics of system evolution as a, e vary. These are still the only two free parameters in the equation, the d is a structural change, not an additional parametrisation.

As the factors in ie are the terms of a complex conjugate pair, and were introduced initially (Gregson, 1988) for that reason, the delay pattern [8.6] with $\omega 5$ is examined first, for varying d. The pattern $\omega 4$ is a sort of mirror image to $\omega 5$; both use a different delay for the a and ae terms. The results of iterating [8.6] for various d show that, within the range of a, e values that admit stability as opposed to explosion, for $j = 1, ..., 40$, effects are relatively simple under the conditions of psychophysical interest. The bifurcation series into quasiperiodicity is preserved and acts as a carrier on which the delay is superimposed; that is, if the evolution of $Y(\mathrm{Re})$ takes the values, say, $Y_1, Y_2, Y_3, Y_4, Y_1, Y_2, Y_3, Y_4...$ for increasing j after the evolution has settled down to a repetitive pattern, by running onto its attractor for those a, e values, then introducing a delay d produces instead a pattern where each Y value is repeated d times; that is in this example $Y_1, Y_1, Y_1, Y_2, Y_2, Y_2, ..$ for $d = 3$. This has the interesting consequence that stopping the process at $j = \eta$ arbitrarily is now more robust in terms of the probability of its Y_{obs} for some imprecision in η. To put this another way, internal delay can have some biological stabilising advantage.

When [8.5] is iterated, however, the repetition breaks down, and more complicated second-order variations within each short run of d terms eventuates. A segment of this process is shown in Table 8.2. One consequence of the introduced delay is an elevation of the $Y(\text{Im})$ values; the system becomes much more noisy, reducing the need to postulate additional random inputs to the model in order to simulate residual variance in response sequences. However, the effects of various different amounts of delay are not simply describable. In the example used for Table 8.2, with the parameters fixed at $a = 3.34, e = .31$, the evolution of Y alters qualitatively as d is increased. When $d = 0$, the trajectory runs rapidly onto the real line at around $Y = (.4516, 0.)$ which is a point attractor. If $d = 1$, then the trajectory runs onto a point which is not on the real line, at around $Y = (.31727, .20295)$. Further increasing d gives the pattern in Table 8.2, which however can explode at later values of j; our interest here as always is only in the early evolution of Y. When $d = 3$, still with the same a, e, the system explodes after wild oscillations, about $j = 25$. But pushing d to 4 also induces explosion, but now not until $j \simeq 40$. All these examples are critically dependent on the a, e settings, and other scenarios will emerge at lower a for varying d. It is not unexpected that such complications are inherent in the system, given the intricacy of the global topology of cubic polynomials, as reviewed in the deep analytic mathematical studies of Branner and Hubbard (1988,1992). It can readily be shown by simulation that the shapes of the Julia sets change drastically as d is increased, and lose their fractal character for $d \geq 3$, but for psychophysical purposes what is more interesting is the change induced in the form of isosimilarity contours, as generated in Chapter 7, which as d increases become more and more constrained in their region and steeper in gradients for high λ, but more stable for low λ.

The point to be emphasised is that delay cannot arbitrarily be entered into the system without consideration of the structure of the Γ expression; when delay, which can be either destructive or preservative of information transmission, is built into a nonlinear system, it has to be entered into mathematical components which are strictly interpretable in their meaning. The substitution in [8.5] might appear to violate this requirement, yet it is dynamically interpretable when considering the trajectory as a time series.

Delay-Induced Transients at the Edge of Stability

The introduction of delay by $\omega 7$, in a peripheral region of the stable Γ parameter space, which shows an almost inverse relation between a and e settings, induces transient dynamics before the ${}^{d}\Gamma_{\omega 7}$ trajectory settles onto a point at-

Table 8.2

Evolution of [8.5] for $d = 2$ from $20 \leq j \leq 40$

Parameters: $a = 3.34, e = 0.31, Y(0) = (0.5, .0001)$

j	Y(Re)	Y(Im)
20	(.534862	-.055822)
21	(.539016	.069068)
22	(.662302	.088254)
23	(.710117	-.019658)
24	(.613366	-.118331)
25	(.487242	-.031246)
26	(.555040	.116595)
27	(.710558	.075014)
28	(.726510	-.069230)
29	(.553854	-.143169)
30	(.445900	.026525)
31	(.602768	.167655)
32	(.751756	.042138)
33	(.726654	-.142172)
34	(.480575	-.139990)
35	(.433540	.116727)
36	(.689942	.204529)
37	(.789066	.014365)
38	(.695258	-.249704)
39	(.391490	-.138469)
40	(.458055	.205440)

tractor. If there is no delay (${}^{d}\Gamma_{\omega 1}$) then the trajectory runs rapidly onto the real line at a point Y_η(Re) proportional to a (Gregson, 1988).

But in the boundary region of stability observed when $d > 0$, which is in the neighbourhood of the Julia set[3] whose structure is a fractal, it is possible

[3] The mathematics of the situation involves no precedents; the interpretation of such dynamics is fully discussed by, for example, Peitgen, Jürgens and Saupe (1992) in their Section 10.8 (p. 575) on the shadowing lemma, in Section 11.5 (p. 647) on chaotic transients, and in Section 12.8 (p. 757) on fractal basin boundaries. They note (p. 769) that in the regions discussed here "Usually, an unending filigree entanglement and increasing bargaining for even the smallest area results".

to have trajectories which oscillate for only a few iterations[4]. The shapes of these transients are interesting because they qualitatively resemble Average Evoked Potentials (AEPs) in electroencephalogram (EEG) recordings from the human scalp when there is sensory input (Begleiter, 1979), with one proviso. That is, the P300 component is not matched. Given that the AEPs can in fact vary considerably in their shape, then the relative magnitudes and the relative latencies of the peaks, considering only the earlier sensory or stimulus-induced components and not the later components which are susceptible to a diversity of cognitive variables, appear to have envelope contours resembling those of the ${}^{d}\Gamma_{\omega 5}$ transients. The analogy is suggestive but too much speculation should not be allowed to rest on it.

The more usual methods of statistically decomposing the shapes of AEPs into components has been based on linear superposition through principal components or fast Fourier transforms. It has also been shown by Başar (1983) that the shape of an AEP can be produced from the dominant EEG frequency components (alpha, beta, theta) by locally varying their relative amplitudes and phase differences. In the present context the approach is fundamentally different, the 'components' are not real dynamic components but are only surface features of the oscillations. There are no bases for decomposition (none are needed!) apart from the $Y(\mathrm{Re})$ and $Y(\mathrm{Im})$ parts, both of which can be chaotic in the transient period, and $Y(\mathrm{Im})$ can persist chaotically afterwards.

Figure 8.1 shows two plots, e on a and $Y(\mathrm{Re})$ on a (the units of the ordinates are fortuitously the same for e and for $Y(\mathrm{Re})$ over the cases considered). The plots are over a sample of points computed; they all lie in the boundary region where the dynamics are only just stable, and liable to be captured by trajectories leading to explosion. The fractal boundary region of the dynamics in the a, e space will lie just to the right of the line drawn, which is almost a straight line in the plot but that is not intrinsically important. The a, e line is bounded in a by the Γ dynamics; it is not extendable outside the range shown in Figure 8.1.

The other line shows the asymptotic $Y(\mathrm{Re})$ value of a point attractor (but with $Y(\mathrm{Im}) > 0$ and aperiodic) which eventuates after the chaotic transient phase. This falls off in value if a is taken below its lower working limit of 2; the process is computable but its $Y(\mathrm{Re})$ asymptote has a local maximum around $a = 2$, of $\simeq .38$. That is, the dynamics in this borderline region strongly reduces

[4] Dynamics such as these which move from one basin of attraction to another via narrow parameter windows are discussed by Wiggins (1992).

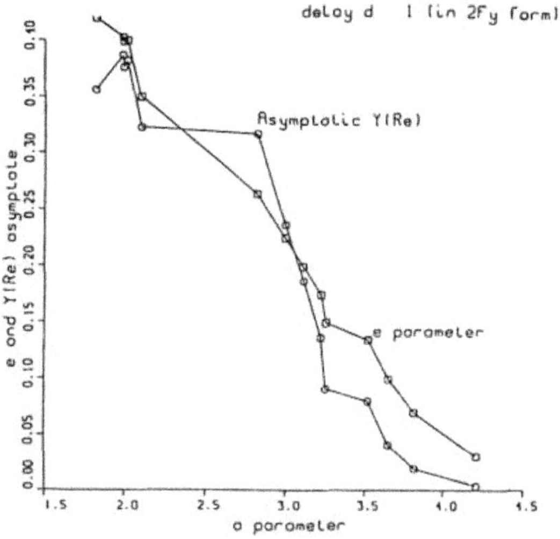

Figure 8.1

Relation beween $a, e, Y(\mathrm{Re})$ in the borderline region inducing transients. Using ${}^{d}\Gamma_{\omega 7}$, i.e. with (1,1,0) on [8.3].

the a input $2 < a < 4$ to an interval $0 < Y_\eta < .38$, for $Y_0 = (.5, \epsilon)$.

The shape of the transient region remains the same for given a, e as a is increased, except that the each peak (negative or positive) is repeated by a multiple d, so that the total number of iterations in the transient is proportional to d. This resembles the property of induced delay [8.6] remarked on earlier. However, in the region where $a \simeq 2$, which corresponds in our previous interpretation to putting a background baseline stimulus level into a channel[5] the shape of the transient oscillation is different. Now there are two or more nearly equal positive peaks instead of one dominant peak. The shape most resembling the typical AEP is only generated with a values (that is, stimulus input levels) in the range $2.8 < a < 4.6$ which is employed for psychophysical simulations. Figures 8.2, 8.3 show examples.

This exploration of delay-induced transients at the boundaries of information transmission raises questions about the transport of the Γ system from

[5] This is the nearest thing corresponding to psychophysiological noise which is used in a signal detection paradigm; but by definition Γ is strictly noiseless, being deterministic.

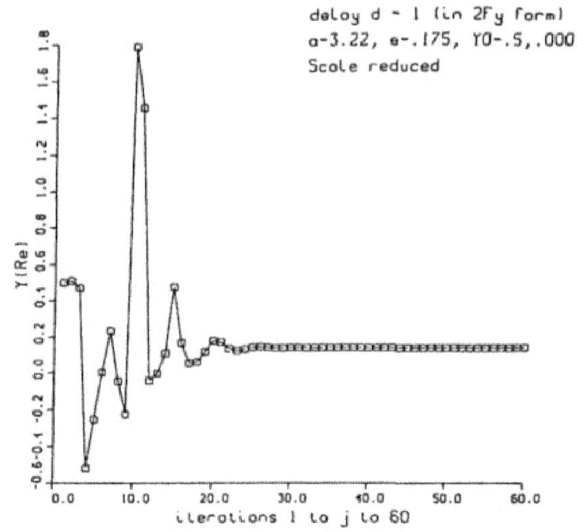

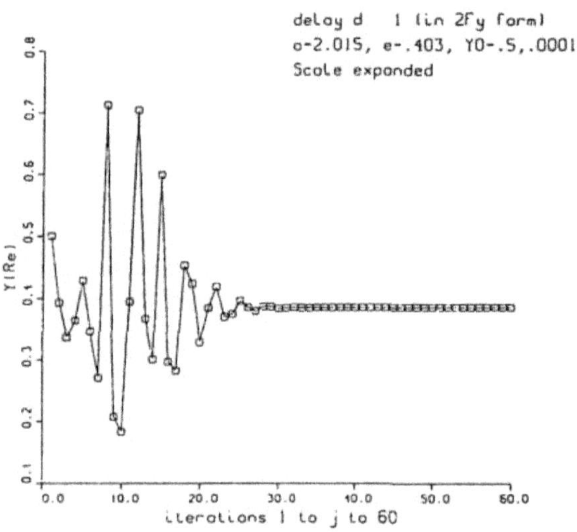

Figures 8.2 and 8.3

Examples of the shape of the transients before stability which can resemble AEPs for stimulus-driven components

background near-chaos into stimulus-controlled responses; the AEP is a physiological artefact from scalp recording, of which neither the subject nor others are ordinarily aware in everyday experience, whereas the $U \rightarrow \Gamma(a,e) \rightarrow Y_{obs}$ mapping is psychological, predicts sensory experience, and has relevance therefore to the content of interpersonal communications. It may be that the two levels of analysis, resting upon the same nonlinear dynamics, relate to the two levels of brain activity that are implicated in AEPs and in sensory consciousness, but this argument cannot be sustained without data that are only now becoming available (Nunez, 1995).

It is important to point out that if we attempt to use a ${}^d\Gamma_{\omega 7}$ process as the core of a cascaded network the results are illuminating but quite unhelpful; images are not preserved but are destroyed. If the process does not explode early, then it becomes chaotic and also creates echoed patterns which replicate multiply, laterally, bearing only a weak relationship to the original input. This finding is consistent with the view that the AEP does not function as a carrier of information, but is a byproduct of information being carried simultaneously at a different dynamical level. Başar (1983) draws attention to the fact that AEPs can appear spontaneously in the background excitation records when there is no information processing of external stimulus signals going on. In the representation explored here, the transients of ${}^d\Gamma_{\omega 7}$ are at the edge of chaos, whereas the functional cascaded networks are mostly in a more stable region of the phase space.

Synchronicity

Features of brain dynamics which could functionally parallel the nonlinear psychophysics are in fact numerous; examples are of two types, either indicating the complexity of hierarchical networked activity, or showing that a lack of one-to-one correspondence betwen brain activity, consciousness, and appropriate discriminatory behaviour, arises[6].

Detection and appropriate action to locate objects can occur to visual stimuli without any conscious awareness, as experiments in the brain-damaged showing *blind-sight* have demonstrated, and of course in the cortically-intact observer there is no conscious access to operations that lead up to the higher

[6] This is no surprise to psychologists who collect data on near-threshold discrimination, or on sensory deprivation; it seems to be a novel idea for the neurophysiologists who have discovered consciousness.

levels implicated in the execution of actions. There is no necessary ground for assuming that all of the transient patterns produced in network evolution, even though they are integral to the dynamics of pattern stabilisation, should have conscious correlates. What is necessary is to show that in the end products of the dynamics are embedded some images that are recognised analogues of human phenomenology.

The evidence from experiments on stabilised images is particularly intersting (Pritchard, Heron and Hebb, 1960). A complete pattern does not fade out simultaneously in all its parts, and lines seem to preserve their integrity and to vanish as a whole. Different sides of a figure can fade at independent rates, and jagged lines are less stable than ones without singularities.

It has been shown in the simulations of Chapter 5 and 6 that some line segments are more vulnerable to distortion than are others. This was suggested as being related to orientation, relative to the lattice grid, as the dominance of vertical and horizontal components in visual illusions is well-known, but intereactions between parts of an extended figure are implicated both in the $(n \times n)\Gamma$ theory and in the stabilised image experiments, both in a dynamic fashion extending through time. Again, though illusions are generated very quickly, they can lack persistence and are not one-shot. Crick (1994, pp. 243-245) uses as a central idea what he calls "binding", in which consciousness is associated with patterns of synchronous oscillation between families of connected neurons. This binding can be long-term or transient; the interlocking of oscillations arises when inputs are being actively processed, though it is not asserted that the temporal onset and termination of binding would be exactly synchronous with the emergence and end of conscious experience. We have shown that synchronous and asynchronous periodicities can be readily produced at different pixels within the same $(n \times n)\Gamma$ lattice, given a simple geometrical figure as input. Thus the nonlinearity does generate internal processes which in their dynamics do superficially resemble some of the corresponding activity in the brain. Whether quantitatively, in terms of frequency, amplitude, phase, and D_2 estimates, there is any but the vaguest resemblance is an open question.

One rare special case, that could not be revealed in perception in the intact normal brain, arises when a patient has no colour sense, for him or her the world is monochromatic, but at the same time can see that two colour patches of different hue which are matched for reflectance can be discriminated if they touch, but can not see that they are different if they are separated by a space. To explore if these two situations have very different internal representations when an $(n \times n)\Gamma$ network is cascaded, four simple input patterns are compared.

These are: (1) a single straight line in the field, (2) a single straight shallow step in the field, (3) a single straight high step in the field, and (4) a single straight shallow step with a thin gap between areas. To contrast these four illustrative cases, sections orthogonally across the step have been drawn for $k = 11, \eta = 5, 2.8 < a <, 4.2, 4.8, \lambda = 1.3$ for all four cases, in Figure 8.4, and for $k = 11, 12, 15$, and $16, \eta = 5, \lambda = 1.3$ for (4) only in Figure 8.5. The sections, as in previous chapters, have been staggered on the ordinate to separate otherwise superimposed plots.

It can be seen from Figure 8.4 that the noise to be unscrambled by the system, upstream from the cascaded lattice, is more complicated if a gap is introduced between areas. The irregularities actually start at the step, and spread out with cascading. How irregular the section is depends on the height of the abrupt step, and on the presence or absence of the tiny gap between areas. A gap does not behave the same way as a thin line on a uniform low intensity field. Another complication is that odd and even cascades do not simply reflect the pattern, and that after about $k = 12$ the sections in (4) can jump to even more qualitative irregularity. As might be expected, the parameter λ is critical here. It is reasonable to conclude that if the system loses processing capacity some patterns will be much more difficult to interpret; quite tiny changes can be important.

From Intrasensory to Intersensory Processes

The phenomena of intersensory interaction, usually weakly facilitative, are summarised by stating that the perception of events in one sensory modality is changed by concurrent stimulation of one or more other sensory modalities.

The existence of such changes appears to be difficult to demonstrate consistently or conclusively. Welch and Warren (1986) provide a valuable review of experimental results up to the mid-1980s, and taken with earlier surveys of work which originated in the Soviet Union (London, 1954), enable us to summarise some empirical findings and list competing theories. It is necessary to mention some of these first, though the present objective is to use a multidimensional psychophysical model, as an abstract algebraic structure, and to examine its properties in direct comparison with some reported phenomena. As might be expected, the psychological theories so far advanced are not in themselves expressed in sufficient detail to be unambiguously mathematised, and ancillary assumptions may need to be introduced.

Various psychological mechanisms have been defined by theorists in terms of the demand characteristics and structure of experimental situations which

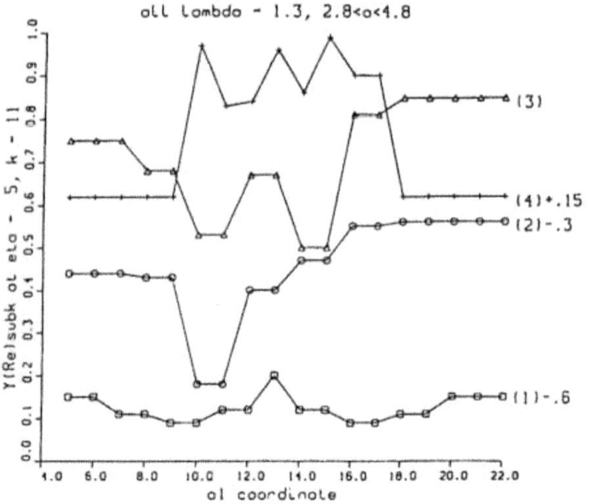

Figure 8.4

The four conditions compared, at $k = 11$, for $Y(\mathrm{Re})$ outputs across the step at $a_1 = 13$.

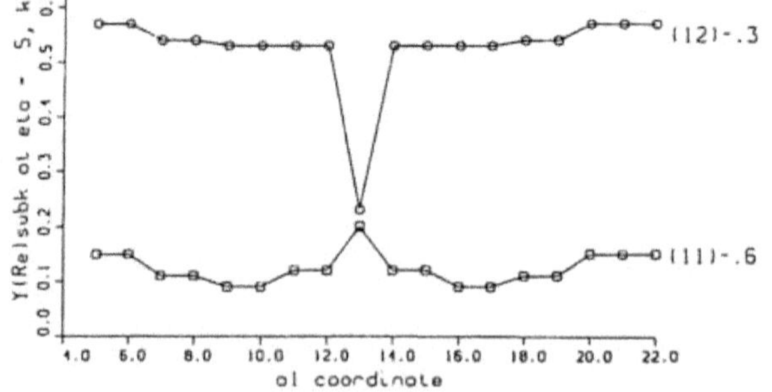

Figure 8.5

Condition (4) with a single step and a thin gap, at cascades $k = 11, 12$, and $15, 16$; evolution of $Y(\mathrm{Re})$.

incorporate multimodal stimulation. Direct intersensory effects have been attributed variously to physiological processes, to attention deployment and distraction, to suggestibility, and to changes in response criteria.

Nonspecific accessory stimulation (the term is Welch and Warren's, and aptly describes much of the earlier Russian work) refers to any second modality sending collateral signals alongside the target modality, and the consequences are expressed in one or more of three ways; changes in detection thresholds of the target modality, changes in response latencies of detection, and changes in perceived sensation intensities. It is clear that processes extending through time, and not instantaneous, are involved, which leads to distinctions being drawn between warning signal effects, where obviously the collateral stimulation precedes the target in time, accessory signal effects, with near simultaneity of channel onset times, and switches of attention between modalities. where both signals are concurrent and temporally sustained.

It has been suggested that intersensory effects might be cognitive rather than sensory, but evidence against this interpretation has gradually accumulated (Pylyshyn, 1980, Radeau, 1992), and the term *cognitive impenetrability* has been used to describe the fact that at least auditory-visual interactions are computationally autonomous in their effect magnitudes, and are not driven by suggestions to, or prior beliefs of, the observer.

When inputs are in two or more modalities simultaneously, a sort of total channel overload can arise, and three kinds of theories have thus been advanced; intersensory bias, (some modalities are more heavily weighted in terms of their inputs) divided attention (the observer switches selectively between inputs) or something called *modality appropriateness*, which refers to some sorts of stimulus variation in space and time being more readily encoded through a particular modality, for example spatial localisation tasks are better done through vision.

It is believed that the resolution of an intersensory discrepancy, where signals conflict in their information or meaning, will always be in favour of the modality which can encode signals more precisely (this sort of precision corresponds to Weber fractions) and so a dominance order, d^*, on modalities will be, with reasonable stimulus levels,

$$\textit{vision } d^* \textit{ proprioception } d^* \textit{ audition } d^* \textit{ chemical senses}$$

Most studies in work reported in English appear to favour vision-audition interactions as more readily amenable to experimental control, visual-chemical senses is a rare pairing (Gregson, 1964).

Welch and Warren (1986, p. 25-23) advance an hypothesis[7] that

> *The greater the number of redundant stimulus properties* (which presupposes some measure of intercoherence of signal strings across modalities) *and the greater the familiarity of the intersensory relationship* (which implies some prior establishment of a bias on cross-coupling parameters between modalities), *then the more unified the intersensory percept will be* (this again implies measures of statistical concordance across time series), *and the greater the contribution of the more appropriate channel will be to the total percept* (there will be various ways of measuring relative contributions of components to a total outcome).

In moving to consider if intersensory effects are a special case of nonlinear psychophysical dynamics, it is proper to ask if variables such as precision, familiarity, appropriateness, are necessary in themselves or can be replaced by derivatives from more primitive operators. It is helpful initially to note that a sensory modality can and normally does support more than one psychophysical dimension (which we will call in its dynamics a *channel*) and that the onset time of channels in a multisensory context is rarely, if ever, realisable in exact temporal synchrony. It follows that in $n\Gamma$ theory the simplest case whose dynamics could be relevantly similar would be 4Γ with stepped inputs. This case has been previously studied briefly as purely intramodal with one cross-coupling parameter λ. With four channels (U_1, U_2, U_3, U_4) in order of onset, the innovation here is that two may be allocated to one hypothetical target modality and the other two to a second accessory source of stimulation. That is, the minimum of two psychophysical channels within each modality is represented.

This 2Γ cross-coupled process, where sensitity in a channel is modified by gain in collateral neighbouring channels, corresponds to the hidden layer in more familiar networks. But the intrinsic mathematics are different from those used in PDP, or for example by Hopfield or Kohonen.

Extension by Relaxation of the Model

In all the $n\Gamma$ and $(n \times n)\Gamma$ theory used so far, λ is a constant parameter over all cross-couplings; this is an oversimplification when any attempt is made to represent the intricacies of intersensory processes, which incidentally are also known to involve hippocampal activity. Varying the cross coupling between

---

[7] Their part is italicised, my interpolations are in parentheses.

channel dynamics[8] in their evolution implies that as soon as the constraint on $\lambda = const$ is relaxed, and a different λ is allowed for each channel, the situation becomes complicated. The model is now $n\Gamma$ **Case 2** with unconstrained interchannel cross-linkage, for brevity this will now be denoted by $n\Gamma_{2,u}$, $n = 4$, the corresponding **Case 1** with no cross-coupling will be $n\Gamma_{1,u}$, $n = 4$. The relations between a, e, λ are summarised for $n\Gamma$ in Table 8.3.

Table 8.3: Interrelation of parameters
in 4Γ recursions
substitutions for the e parameter in each Γ

Channel	$n\Gamma_1$	$n\Gamma_{1,u}$	$n\Gamma_2$	$n\Gamma_{2,u}$
U_1, a_1	e	e_1	$\lambda/a_{m,c}$	$\lambda_1/a_{m,c}$
U_2, a_2	e	e_2	$\lambda/a_{m,c}$	$\lambda_2/a_{m,c}$
U_3, a_3	e	e_3	$\lambda/a_{m,c}$	$\lambda_3/a_{m,c}$
U_4, a_4	e	e_4	$\lambda/a_{m,c}$	$\lambda_4/a_{m,c}$

In Table 8.3 all terms are substitutions for the second parameter e. The notation $a_{m,c} = max\{S_c\}$, where for $c = 1, 2, ..., n$ $S_c = \{\{a\} - a_c\}$, is needed.

It can be deduced from that under some restricted input patterns $\{U\}$ the constrained and unconstrained models with yield the same $U \mapsto Y(\text{Re})$ for a given set $\{a\}$. This is the situation called *compensation* (Gregson, 1992a).

$4\Gamma_{2,u}$: The most unrestricted case considered; the logic is the same as the previous case, but the restriction on λ is lifted, so that the total $(2n)$ parameter set is now $a_1, a_2, a_3, a_4, \lambda_1, \lambda_2, \lambda_3, \lambda_4$. The effect of fixed λ in $n\Gamma_2$ is to push the highest channel, the one with $a = a_{m,c}$, into the upper part of the psychophysical ogive for $a \mapsto Y(\text{Re})$ and the remaining channels to some degree into the lower part. This is loosely analogous to pushing the system onto a mapping resembling the supremum metric; the dominance of the largest input is always enhanced. This can happen also for some parameter settings in $n\Gamma_{2,u}$, but not invariantly. Under some restricted and specific settings of the parameters, this case can be indistinguishable in its input-output mappings from $n\Gamma_{1,u}$.

[8] Some of this section was first reported in an invited address to the 1995 Neural Networks Conference at the University of Sydney, NSW.

An Introductory Example

There are various alternative predictions that can be derived from the evolution of the trajectories in $4\Gamma_{2,u}$; using all the available theory, the mean and s.d. of outputs Y, and the mean and s.d. of latencies in j (recursion) steps, and the terminal values $Y_\eta(\text{Re}) = Y_{obs}$ are all computable. The $meanY$ correspond approximately to values fitted in traditional psychophysics where equations such as Fechner's or Plateau's are used. For some quite arbitrary a values, drawn from 3.72, 3.74, 3.76, 3.78, 4.72, 4.74, 4.76, 4.78, an illustrative example is summarised in Table 8.4. This simulation makes each input a rectangular a pulse of a short duration $\not> 10j$, the plotted trajectories are thus destabilisation and restabilisation curves (see Chapter 3) in immediate succession for each channel, with the onset and termination points slightly staggered between channels. The first point to notice is that the nonlinearity of the system creates intricate patterns with the interplay of the order of input, and the λ and a values used. The terminal $Y(\text{Re})$ values are not always in the same rank order as the means ($meanY$s in Tables 8.4-8.6) and the trajectories of channels cross over one another during the system's evolution to $\eta = 20$. This is shown in the four plots of trajectories, Figures 8.6 through 8.9. It is also possible to plot the convolutions of the Y_j and the probabilities of termination of the evolution at any j using the derivative of the Jacobian of the evolution; this is used to obtain $meanY$ and the mean latencies shown in the tables. For the methods see Gregson (1992a).

The $\sum Y$ terms correspond to the total sensation intensity, obtained by summing across channels at each iteration. These results may also be needed to compare with other intersensory phenomena.

The trajectories are each for an internal component channel; they rise with destabilisation of the background, and restabilise. They could equally well be computed to drop out of chaos into limit cycling, but as shown they rise from point attractor dynamics into limit cycles, or quasiperiodicity, transiently. Each curve shows peak values, or peak oscillations, and restabilisation levels which are not necessarily the same as starting levels. The rates of rising and falling vary across channels. From the trajectories, and their associated properties, one can compute latency distributions, and from convolutions of latency and Y values, mean values for an external observer, are derivable. In Table 8.4, which matches the Figures, the conditions all involve the starting points of each channel being staggered, so that Dimn. 1 precedes Dimn. 2 by one j step, and so on. The a values are nearly the same for the first two inputs in Settings

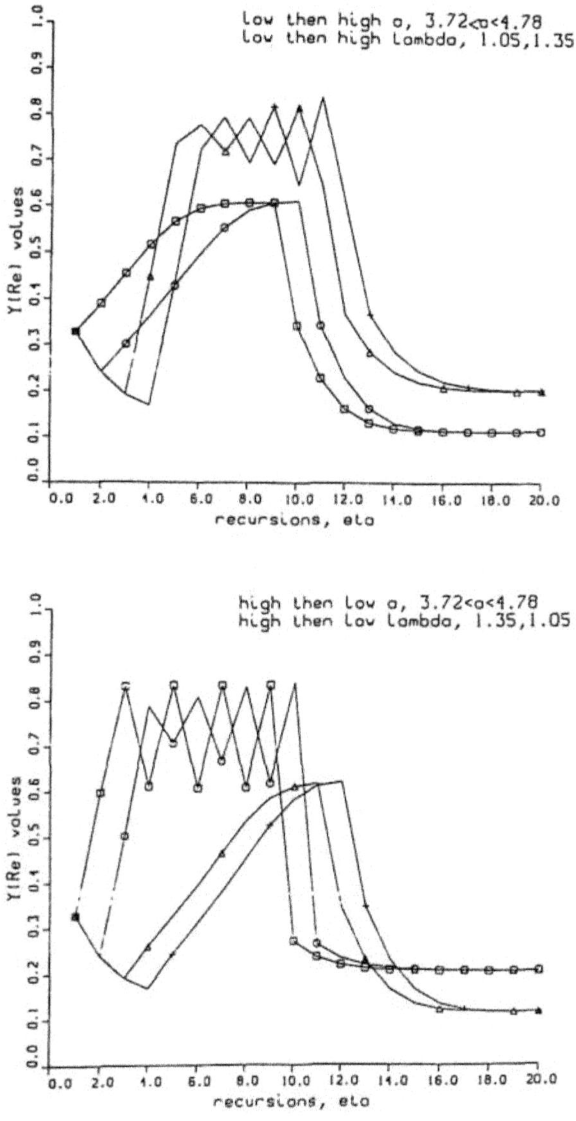

Figures 8.6 and 8.7

Trajectories for Settings 1 and 2 in Table 8.4

1 through 4, (which match the Figures 8.6 through 8.9) and nearly the same for the second two inputs but different from the first two. In the Settings 5 and 6 (not graphed, not so interesting) the a values are the same throughout. The λ values are treated in the same way, so for example for Setting 1 the sequence of a values is 3.72,3.74,4.76,4.78, and the λ values are 1.05,1.05,1.35,1.35.

The effects of interdimensional interaction depend in this example partly on order of inputs, and partly on parameter settings, both of gains a and of cross coupings λ. In some cases the presence of preceding low inputs does nothing to the mean or terminal values of high inputs; compare in Table 8.4 Dimensions 3 and 4 in Settings 1 and 5. On the other hand, Dimensions 3 and 4 in Settings 2 and 6 tell a different story; preceding high inputs have a suppressive effect on both mean and terminal Y(Re) values. The peak values of each trajectory are not necessarily in the same order of magnitude across dimensions as are the terminal values at Y_η; this is easiest seen by comparing graphs.

Intersensory Induction Effects

To model intersensory effects such as facilitation, it is necessary to compare different parameter settings, using cases where some channels are set with $a = a_0$. An example is worked through in Tables 8.5 and 8.6; this treatment is in effect using the a_0 conditions as a control, with two dimensions acting as dummy information carriers to hold the total system dimensionality constant over comparisions. The same framework as used in Table 8.4 is employed, but now a comparison is made between outputs which result from two parameter settings for the total $4 - d$ system.

To model small intersensory facilitative effects, two stages are needed; we first explore with the heuristics of Tables 8.5 and 8.6, to see if there exists a domain within a setting that exemplifies what is suspected to happen. Then if this is so, we expand the parameter space in the neighbourhood of a result of interest to find a parametric description of the conditions where consistent facilitation can occur. Remember always that facilitation is rarer than masking, so it may need searching for in the parameter space.

In Table 8.6, for Panels 3 and 4, Target dimension 2, (the first in order to be input), has as a shift in $meanY$ from .345 to .350 when the collateral channel is lifted markedly from zero, and the λ values for target and collateral are as shown in the lower half of the table. The associated $s.d.Y$ is reduced from .137

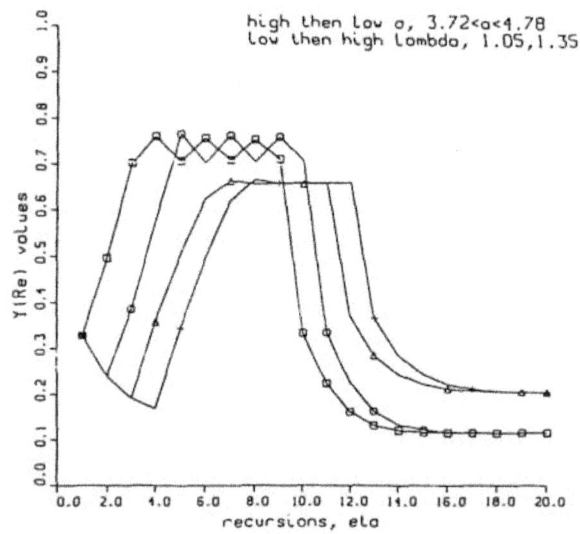

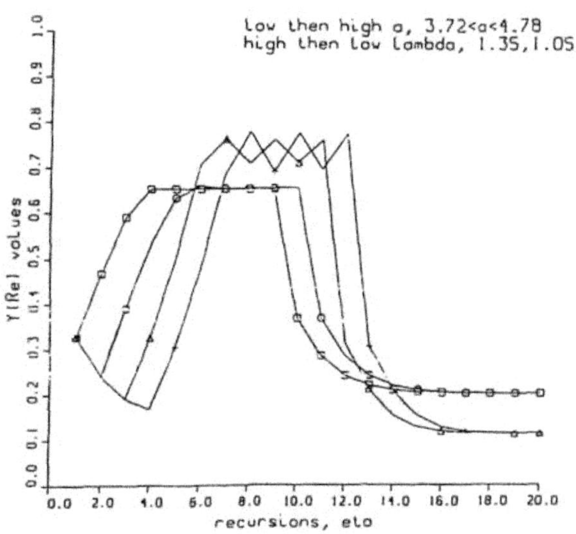

Figures 8.8 and 8.9

Trajectories for Settings 3 and 4 in Table 8.4

to .129, the latency is reduced from $4.01j$ to $3.86j$, and its s.d. correspondingly drops. But the total Y is also reduced.

The means the simulated phenomenon is weak, even if consistent in all its parameter shifts, but could vary if responses are delayed. The predictions for the first stage of the simulation analysis are that the intersensory facilitation of a target channel depends on that channel being initial within the target modality, and that the λ parameters have to be set so that the collateral inputs whose activity precedes the onset of the target stimulation, have to be more resistant to intersensory inhibition than the target. It is as though the context drags the target levels upwards.

From the outside of the Black Box, these dynamics would look like facilitation, shortening of response latencies, and improvement in response precision and the estimation of target intensity. Increase in precision is analogous to threshold suppression, or to a change in the gradient of the psychophysical function; that is modifiable by e (and hence λ) which is the single channel sensitivity. But all this happens in a narrow parameter region, critically dependent on the time interval between target and collateral stimulus onset. It takes some experimental exploration to find such a region in a real system, which could account for the inconsistency of different reported studies, and the advancement of special theories about attention or about nonlinear physiological masking.

It should be noted that there are effects in the opposite direction which can be quite marked. Consider for example a reinterpretation of the channels 1 and 4 results in panels 1 and 2 of Table 8.5, if the meaning of target and collateral is reversed so that 1,4 becomes target and 2,3 becomes a collateral pulse during a longer target presentation. This is like a brief switching of attention. We now have, with this reversal of interpretation, zero target level becoming high target level, and the results are quite simple; when the target a increases the $meanY$ and associated properties are shifted compatibly. Here the higher target activity suppresses the collateral pulse (channels 2 and 3) $meanY$ levels slightly.

What is not usually explored experimentally is the effect of the target on the collateral stimulation; in other words, reversing the sense of the judgments part way through a series of stimulus presentations.

Suppose we have again the case reported in panels 3 and 4 of Table 8.6, and elicit judgments on channel 4 only. This is then a replication of the case just considered, output goes up with input monotonically.

Another important comparison is made by holding the target – collateral interpretation as given originally, and matching the 2nd and 3rd channels across

Table 8.4: Interdimensional Interactions

Example with 4Γ outputs, high $\lambda = 1.35$, low $\lambda = 1.05$

Dimn:	mean Y	s.d. Y	mean Laty	s.d Laty	Y_η(Re)
		Setting 1 low-high a, low-high λ			
1:	.478	.097	4.14	2.71	.114
2:	.358	.126	4.09	2.37	.114
3:	.552	.250	5.09	2.28	.202
4:	.480	.239	5.37	2.03	.205
$\sum$:	1.881	.742	5.11	3.13	.635
		Setting 2 high-low a, high-low λ			
1:	.710	.129	4.35	2.16	.204
2:	.546	.253	4.87	2.12	.206
3:	.299	.129	4.38	2.21	.115
4:	.253	.122	4.68	2.11	.115
$\sum$:	2.343	.160	7.10	2.61	.640
		Setting 3 high-low a, low-high λ			
1:	.681	.126	4.42	2.49	.115
2:	.487	.214	4.24	2.35	.116
3:	.467	.225	5.57	3.10	.204
4:	.402	.226	5.56	3.02	.205
$\sum$:	2.327	.593	6.19	3.11	.640
		Setting 4 low-high a, high-low λ			
1:	.586	.111	6.32	3.00	.202
2:	.477	.204	5.67	3.16	.202
3:	.449	.215	4.46	2.19	.114
4:	.340	.231	4.81	2.07	.115
$\sum$:	2.017	.703	5.37	3.14	.633
		Setting 5 high a, high λ			
1:	.710	.127	4.36	2.17	.202
2:	.535	.250	4.93	2.22	.202
3:	.552	.250	5.08	2.28	.202
4:	.480	.239	5.37	2.02	.205
$\sum$:	2.521	.441	5.98	2.45	.811
		Setting 6 low a, low λ			
1:	.580	.111	6.17	3.04	.195
2:	.471	.202	5.53	3.15	.195
3:	.446	.225	5.41	3.06	.195
4:	.389	.222	5.45	2.98	.198
$\sum$:	2.044	.620	5.93	3.15	.781

Table 8.5: Interdimensional Interactions

Target dimensions are 2 and 3; $a = 3.76$, $\lambda = 1.35$
Collateral dimensions are 1 and 4; $\lambda = 0.85$

Panels 1 and 2:

Dimn:	mean Y	s.d. Y	mean Laty	s.d Laty	Y_η(Re)
	Setting with zero collateral channel $a = 2.03$				
1:	.129	.029	5.54	2.88	.114
2:	.378	.130	6.68	2.89	.313
3:	.369	.127	6.55	3.00	.313
4:	.135	.035	5.41	2.89	.114
$\sum$:	.928	.154	6.79	3.05	.928
	Setting with high collateral channel $a = 4.03$				
1:	.348	.155	3.89	2.85	.098
2:	.331	.101	6.55	3.01	.273
3:	.329	.105	6.35	3.09	.273
4:	.314	.122	4.38	1.95	.098
$\sum$:	1.161	.472	5.74	3.23	.742

panels. This in fact makes a big difference for the parameter values used. If λ for the target channels is much higher than for the collateral channels, then the contrast between $meanY$ values (channels 2 and 3 versus 1 and 4) is biggest in the first panel. The shift is in the opposite direction if the 2nd and 4th panels are compared, which would be represented as an interaction effect if an attempt were made to use Manova to disentangle a full factored experimental design. The higher λ setting for the target channels apparently corresponds to what was called *modality appropriateness* by Welch and Warren (1986).

Expansion of the exploration of the parameter space can now proceed; the independent variables of interest are $\lambda_{1,4}$ and $\lambda_{2,3}$. The dependent variables are $meanY_2|a_{1,4} = 2.01$, and $meanY_2|a_{1,4} = 4.01$. As a are fixed, this situation can be reduced to one dependent variable ΔY, but the bivariate plot of ΔY on $\lambda_{target} \times \lambda_{collateral}$ has to be explored as a contour map as in Figure 8.10. This turns out to be rather odd; the shift in the target Y is independent of λ(collateral) but changes from positive to negative to positive as λ(target) is increased.

Table 8.6: Interdimensional Interactions

Target dimensions are 2 and 3; $a = 3.76$, $\lambda = 0.85$
Collateral dimensions are 1 and 4; $\lambda = 1.35$

Panels 3 and 4:

Dimn:	mean Y	s.d. Y	mean Laty	s.d Laty	Y_η(Re)
	Setting with zero collateral channel $a = 2.03$				
1:	.316	.002	5.97	2.74	.313
2:	.345	.137	4.01	2.38	.114
3:	.333	.133	4.30	2.17	.114
4:	.298	.041	6.24	2.63	.313
$\sum$:	1.239	.342	5.14	2.66	.855
	Setting with high collateral channel $a = 4.03$				
1:	.435	.211	6.26	2.79	.273
2:	.350	.129	3.86	2.05	.098
3:	.333	.124	4.31	1.88	.098
4:	.388	.175	6.12	2.92	.273
$\sum$:	1.268	.532	5.84	3.01	.742

The expansion of the comparison between panels 2 and 4 in Tables 8.5 and 8.6 is made by varying $a_{1,4}$(collateral) against fixed $a_{2,3}$(target) with the two λ settings employed. The dependent variables are $\Delta meanY_2$ and $\Delta meanY_3$ so two plots are required; $\Delta meanY_2/a_{1,4}$ and $\Delta meanY_3/a_{1,4}$ in Figures 8.11 and 8.12.

The comparison between Figures 8.11 and 8.12 generates the predictions about the effects of target onset delay on relative judgments of intensity. The settings for $a_{1,4} = a_{2,3}$ (one line at 45° to the axes, which is virtual on the Figures) are a homogeneity condition, there only λ effects are operative. The small differences between Figures 8.11 and 8.12 are due to order effects between the stepped inputs of the four channels. The two target channels have the same parameters (a and λ) but their evolution is sufficiently different to make some small changes in their trajectories. It is unlikely that a real experiment could reliably identify these differences if tested within a statistical model that has H_0 : $(meanY_2 = meanY_3)$. It is possible to argue that intermodal and intramodal interactions may be mediated by the same nonlinear multichannel

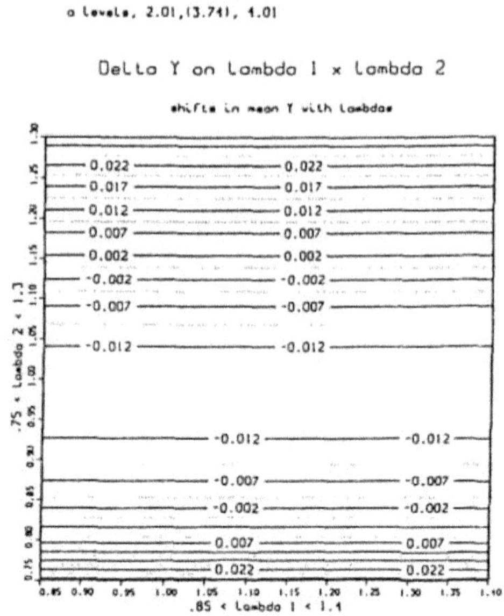

Figure 8.10

processes but with different internal parameter settings.

Inevitably the creation of a 4-d system with nonlinear dynamics and cross-coupling betwen channels is expected to create complications in input-output mappings; the interplay of the parameter settings, over time, can yield maps which vary from the simplest linear or quadratic (Figure 8.10) to the surfaces (Figures 8.11 and 8.12) which are irreducible to a linear system satisfying the superposition principle. Only within a very local region of the parameter space can piecewise linearisation be useful.

Conclusions

At the start of this monograph, in Chapter 1, some similarities and differences between three-layer neural networks and $n\Gamma$ (but not, in fact $(n \times n)\Gamma$) mappings were noted, mainly to arrest the reader's temptation to see analogies too readily. It is worth noting, therefore, that backpropagation is a core feature of parallel distributed processing in cognitive psychology, but does not occur in real brains and does not occur in the dissipative systems created in our cascades. Admittedly, the systems used here to create a sensory psychophysics, and extend to apperception, by parallel activities recursively serialised, is not

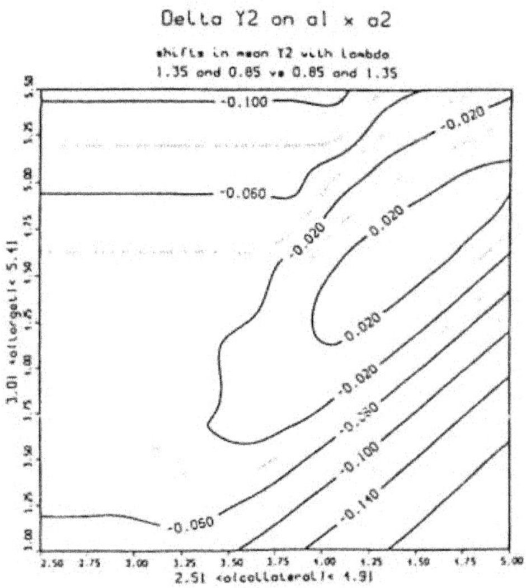

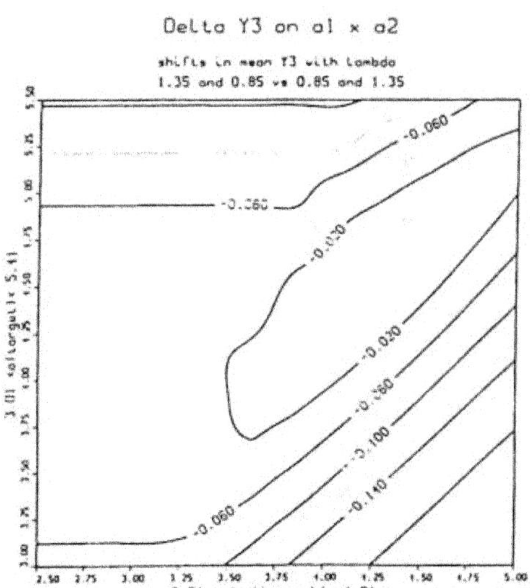

Figures 8.11 and 8.12

about learning or memory, and backpropagation was an AI device to create learning capacity. The contrast in this comment is about the sequence of transformations in the internal variables of Γ trajectories, and the global dynamics of brain activity specifically associated with the edge of consciousness. Memory is not conscious, its reconstructions are brought into consciousness through pathways that need not resemble those featuring immediate sensory awareness. Still, contemporary discussions add considerably to fine detail, but little in principle, to what Fechner, when creating psychophysics, actually wrote[9];

> "... mag der Gedanke am Flusse der körperlichen Thätigkeit selbst mitwirken und nur mittels dieses Wirkens wirklich sein, oder mag er des Flusses nur bedürfen, wie der Ruderer im Nachen, um darüber hin zu steuern, und dabei mit dem Ruder gleichgültige Wellen schlagen, beidesfalls wollen die Verhältnisse und Gesetze des Flusses berücksichtigt sein, wenn es sich um den Fluß oder Fortschritt des Gedankens handelt".
> G. T. Fechner, *Elemente der Psychophysik,* 1860, Bd. 1, S. 14.

---

[9] Trans: ..thought may be considered as itself participating in the stream of bodily processes, and actually only becoming a reality by means of these processes, or it may only need this stream in the way that an oarsman steers the boat, and goes across, making insignificant ripples with his oar, yet at the same time the conditions and laws of the river's motion must be taken into account when the flow or progress of thought is concerned."

References

Aitchison, J. and Shen, S. M. (1980) Logistic-normal distributions: some properties and uses. *Biometrika, 67,* 261-272.

Aitkin, M. (1991) Posterior Bayes Factors. *Journal of the Royal Statistical Society, Series B, 53,* 111-142.

Akaike, H. (1985) Prediction and Entropy. *In* Atkinson, A. C. and Fienberg, S. E. (Eds.) *A Celebration of Statistics.* New York: Springer-Verlag.

Allan, L. G., and Kristofferson, A. B. (1974) Psychophysical theories of duration discrimination. *Perception and Psychophysics, 16,* 26-34.

Allan, L. G. (1979) The perception of time. *Perception and Psychophysics, 26,* 340-354.

Allan, L. G. (1992) The internal clock revisited. *In* Macar, F., Pouthas, V. and Friedman, W. J. (1992) *Time, Action and Cognition.* Dordrecht: Kluwer, pp. 191-202.

Altman, M. (1986) *A Unified Theory of Nonlinear Operator and Evolution Equations with Applications.* New York: Marcel Dekker.

Anstis, S. and Saida, S. (1985) Adaptation to Auditory Streaming of Frequency-Modulated Tones. *Journal of Experimental Psychology: Human Perception and Performance, 11,* 257-271.

Arnheim, R. (1960,1974)*Art and Visual Perception.*[First and Second Revised Editions] London: Faber and Faber.

Asch, S. H. and Witkin, H. A. (1948) Studies in space orietation: II. Perception of the upright with displaced visual fields and with the body tilted. *Journal of Experimental Psychology, 38,* 455-477.

Ashby, F. G. and Lee, W. W. (1991) Predicting Similarity and Categorization from Indentification. *Journal of Experimental Psychology: General, 120,* 150-172.

Baird, J.C. and Noma, E. (1978) *Fundamentals of Scaling and Psychophysics.* New York: Wiley.

Barndorff-Nielsen, O. E., Jensen, J. L. and Kendall, W. S. (1993) *Networks and Chaos: Statistical and Probabilistic Aspects.* London: Chapman and Hall.

Başar, E. Synergetics of Neuronal Populations. A Survey on Experiments. *In* Başar, E, Flohr, H., Haken, H. and Mandell, A. J. (Eds.) (1983) *Synergetics of the Brain.* Berlin: Springer-Verlag. pp. 183-200.

Bayer, C. A. and Pressey, A. W. (1972) Geometric illusions as a function of pigmentation of the Fundus oculi and target size. *Psychonomic Science, 26,* 77-79.

Beavois, M. W. and Meddis, R. (1991) A Computer Model of Auditory Stream Segregation. *Quarterly Journal of Experimental Psychology, 43A,* 517-542.

Békésy, G. von (1959) Synchronism of Neural Discharges and their Demultiplication in Pitch Perception on the Skin and in Hearing. *Journal of the Acoustical Society of America, 31,* 338-399.

Begleiter, H. (Ed.) (1979) *Evoked Brain Potentials and Behavior.*New York: Plenum Press.

Benussi,V. (1916) Versuche zur Analyse Taktil-erweckter Scheinbewegungen. *Archiv für die gesamte Psychologie, 36,* 59-135.

Berglund, B., Berglund, U. and Ekman, G. (1967) Temporal Integration of Vibrotactile Stimulation. *Perceptual and Motor Skills, 25,* 549-560.

Bertelson, P. (1966) Central Intermittency Twenty Years later. *Quarterly Journal of Experimental Psychology, 18,* 153-163.

Binet, A. (1895) La mèsure des illusions visuelles chez les enfants. *Revue Philosophique, 40,* 11-25.

Bloch, D. A. and Watson, G. S. (1967) A Bayesian study of the multinomial distribution. *Annals of Mathematical Statisitics, 38,* 1423-1435.

Block, L. S. and Coppel, W. A. (1986) Stratification of continuous maps of an interval. *Transactions of the American Mathematical Society, 297,* 587-604.

Block, L. S. and Coppel, W. A. (1991)*Dynamics in One Dimension. Lecture Notes in Mathematics, # 1513.* Berlin: Springer-Verlag.

Block, R. A. (1990) *Cognitive Models of Psychological Time.* Hillsdale, NJ: Lawrence Erlbaum Associates.

Bock, G. R. and Marsh, J. (Eds.) (1993) *Experimental and Theoretical Studies of Consciousness. Ciba Foundation Symposium 174.* Chichester: Wiley.

Box, G. E. P. and Tiao, G. C. (1965) A change in level of a stationary time series. *Biometrika, 52,* 181-192.

Branner, B. and Hubbard, J. H. (1988) The iteration of cubic polynomials Part I: The global topology of parameter space. *Acta Mathematica, 160,* 143-206.

Branner, B. and Hubbard, J. H. (1992) The iteration of cubic polynomials Part II: Patterns and parapatterns. *Acta Mathematica, 169,* 229-325.

Bregman, A. S. (1978) Auditory Streaming is Cumulative. *Journal of Experimental Psychology: Human Perception and Performance, 4,* 380-387.

Bregman, A. S. and Tougas, Y. (1989) Propagation of constraints in auditory organization. *Perception and Psychophysics, 46,* 395-396.

Broderson, U. (1968) *Intra- und interindividuelle mehrdimensionale Skalierung eines nach objektiven Kriterien variierten Reizmaterials.* Unpublished thesis: University of Kiel.

Brown, S. W., Stubbs, D. A. and West, A. N. (1992) Attention, Multiple Timing, and Psychophysical Scaling of Temporal Judgments. *In* Macar, F., Pouthas, V. and Friedman, W. J. (1992) *Time, Action and Cognition.* Dordrecht: Kluwer, pp. 129-140.

Caelli, T. (1984) Coding Principles for Visual Image Processing. *In* Dodwell, P. C. and Caelli, T. (1984) *Figural Synthesis.* Hillsdale, NJ.: L. Erlbaum Associates. Chapter 5.

Calogero, F. and Degasperis, A. (1982) *Spectral transforms and solitons: tools to solve and investigate nonlinear evolution equations. Vol. I.* Amsterdam: North Holland.

Camacho, C. and Camacho, M. I. (1993) Complex Foliations Arising from Polynomial Differential Equations. *In* Schlomiuk, D. (Ed.) *Bifurcations and Periodic Orbits of Vector Fields.* Dordrecht: Kluwer. pp. 1-18.

Campbell, E. A. and Gregson, R. A. M. (1990) Julia Sets for the Gamma Recursion in Nonlinear Psychophysics. *Acta Applicandae Mathematicae, 20,* 177 - 188.

Chaghaghi, F. S. (1985) *Time Series Package (TSPACK). (Lecture Notes in Computer Science, 187.* Berlin: Springer-Verlag.

Chapeau-Blondeau, F. and Chauvet, G. (1992) Stable, Oscillatory and Chaotic Regimes in the Dynamics of Small Neural Networks with Delay. *Neural Networks, 5,* 735-743.

Ciocca, V. and Bregman, W. S. (1987) Perceived continuity of gliding and steady-state tones through interrupting noise. *Perception and Psychophysics, 42,* 476-484.

Coren, S. (1970) Lateral Inhibition and Geometrical Illusions. *Quarterly Journal of Experimental Psychology, 22,* 274-278.

Coren, S. and Girgus, J. S. (1972) Differentiation and decrement in the Müller-Lyer illusion. *Perception and Psychophysics, 12,* 466-470.

Coren, S., Ward, L. M., Porac, C. and Fraser, R. (1978) The effect of optical blur on visual-geometric illusions. *Bulletin of the Psychonomic Society, 11,* 390-392.

Creelman, C. D. (1962) Human discrimination of auditory discrimination. *Journal of the Acoustical Society of America, 34,* 582-593.

Crick, F. (1993) *The Astonishing Hypothesis: The Scientific Search for the Soul.* New York: Scribner.

Curtis, D. W., and Rule, S. J. (1977) Judgment of duration relations: Simultaneous and sequential presentation. *Perception and Psychophysics, 22,* 578-584.

Dale, A. I. (1991) *A History of Inverse Probability from Thomas Bayes to Karl Pearson.* New York: Springer-Verlag.

Davies, A. R. and Anderssen, R. S. (1986) Optimisation in the regularisation of ill-posed problems. *Journal of the Australian Mathematical Society, B., 28,* 114-133.

Day, R. H. (1994) Apparent Misalignmwent and Misbisection Without Delineated Angles. *Proceedings of the 21st Experimental Psychology Conference, Sydney, NSW.* p. 40 (abstract).

Day, R. H. and Dickinson, R. G. (1976) Apparent length of the arms of acute and obtuse angles, and the components of the Müller-Lyer illusion. *Australian Journal of Psychology, 28,* 137-148.

De Groot, M. H. (1970) *Optimal Statistical Decisions.* New York: McGraw Hill.

Dehaene, S. (1993) Temporal Oscillations in Human Perception. *Psychological Science, 4,* 264-270.

Delsuc, M. A. (1989)A new Maximum Entropy Processing Algorithm with Applications to Nuclear Magnetic Resonance Experiments. *In* Skilling, J. *Maximum Entropy and Bayesian Methods.* Dordrecht: Kluwer. pp 285-290.

de Melo, W. and van Strien, S. (1993) *One-Dimensional Dynamics.* Berlin: Springer-Verlag.

Dodwell, P. C. and Caelli, T. (1984) *Figural Synthesis.* Hillsdale, NJ.: L. Erlbaum Associates.

Donkin, W. F. (1851) On certain questions relating to the theory of probabilities: in three parts. *The London, Edinburgh, and Dublin Philosophical Magazine and Journal of Science, Series F, 1,* 353-368, 458-466, 2, 55-60.

Donoho, D. L. (1988) One-sided inference about functionals of a density. *Annals of Statistics, 16,* 1390-1420.

Donoho, D. L., Johnstone, I. M., Hoch, J. C. and Stern, A. S. (1992) Maximum Entropy and the Nearly Black Object. *Journal of the Royal Statistical Society, Series B, 54*, 41-81.

Eisler, A. D. (1993) *Time Perception: theoretical considerations and empirical studies of the influence of gender, age, and culture on subjective duration.* University of Stockholm: Unpublished Ph.D. thesis.

Eisler, H. (1981) Applicability of the parallel-clock model to duration discrimination. *Perception and Psychophysics, 29,* 225-223.

Eisler, H. (1989) Data-Equivalent Models in Psychophysics: Examples and Reflections. *In* Ljunggren, G. and Dornic, S. (Eds.)*Psychophysics in Action.* Berlin: Springer-Verlag. pp 11-24.

Eisler, H. and Ekman, G. A. (1959) A mechanism of subjective similarity. *Nordisk Psychologi, 11,* 1 - 10.

Eisler, H. and Lindman, R. (1990) Representations of Dimensional Models of Similarity. *In* Geissler, H.-G., Müller, M. H. and Prinz, W. (Eds.) *Psychophysical Explorations of Mental Structures.* Göttingen: Hogrefe and Huber. pp. 165-171.

Ekman, G., Berglund, B., Berglund, U. and Lindvall, T. (1967) Perceived Intensity of Odor as a Function of Time of Adaptation. *Scandinavian Journal of Psychology, 8,* 177-186.

Engelbrecht,J. K., Fridman, V. E. and Pelinovski, E. N. (1988) *Nonlinear Evolution Equations.* Harlow, Essex: Longman.

Erickson, G. J. and Smith, C. R. (Eds.) (1988) *Maximum Entropy and Bayesian Methods in Science.* Boston: Kluwer.

Ferrar, W. L. (1950) *Higher Algebra.* Oxford: University Press.

Foley, J. M. and McCourt, M. E. (1985) Visual grating induction. *Journal of the Optical Society of America, A2,* 1220-1230.

Franzén, O., Markowitz, J. and Swets, J. A. (1970) Spatially-limited Attention to Vibrotactile Stimulation. *Perception and Psychophysics, 7,* 193-196.

Geldard, F. A. (1975)*Sensory Saltation: Metastability in the Perceptual World.* Hillsdale, NJ: L. Erlbaum Associates.

Gibbon, J., Church, R. M. and Meck,W. H. (1984) Scalar Timing in Memory. *In* Gibbon, J. and Allan, L. G. (Eds.) *Annals of the New York Academy of Sciences: Timing and Time Perception,* (Vol 423, pp. 52-77.) New York: New York Academy of Sciences.

Gilden, D. L., Schmuckler, M. A. and Clayton, K. (1993) The Perception of Natural Contours. *Psychological Review, 100,* 460-478.

Globus, G. G. (1992) Towards a noncomputational cognitive neuroscience. *Journal of Cognitive Neuroscience, 4,* 299-310.

Goff, G. D. (1967) Differential Discrimination of Frequency of Cutaneous Mechanical Vibration. *Journal of Experimental Psychology, 74,* 294-299.

Gombrich, E. H. (1960) *Art and Ilusion.* London: Phaidon Press.

Good, I. J., Holtzmann, G. I., Deaton, M. L. and Bernstein, L. H. (1989) Diagnosis of heart attack from two enzyme measurements by means of bivariate probability density estimations: statistical details. *Journal of Statistical and Computational Simulation, 32,* 68-76.

Grandy, W. T. and Schick, L. H. (Eds.) (1991) *Maximum Entropy and Bayesian Methods.* Boston: Kluwer.

Gregson, R. A. M. (1964) Modification of perceived relative intensities of acid tastes by ambient illumination changes. *Australian Journal of Psychology, 16,* 190-199.

Gregson, R. A. M. (1975) *Psychometrics of Similarity.* New York: Academic Press.

Gregson, R. A. M. (1976) A comparative study of seven similarity models. *British Journal of Mathematical and Statistical Psychology, 29,* 139 - 156.

Gregson, R. A. M. (1980) Model evaluation via stochastic parameter convergence as on-line system identification. *British Journal of Mathematical and Statistical Psychology, 33,* 17 - 35.

Gregson, R. A. M. (1983) *Time Series in Psychology.* Hillsdale, NJ.: L. Erlbaum Associates.

Gregson, R. A. M. (1985) Vergleich einiger mengentheoretischer und distanz-Repräsentation der Änlichkeitsdaten von Broderson (1968). *Zeitschrift für experimentelle und angewandte Psychologie, 32,* 573 - 587.

Gregson, R. A. M. (1988) *Nonlinear Psychophysical Dynamics.* Hillsdale, NJ.: L. Erlbaum Associates.

Gregson, R. A. M. (1989) A nonlinear systems approach to Fechner's paradox. *Biological Cybernetics, 61,* 129 - 138.

Gregson, R. A. M. (1992a) *n-Dimensional Nonlinear Psychophysics.* Hillsdale, NJ.: L. Erlbaum Associates.

Gregson, R. A. M. (1992b) The Psychophysical Method of Limits; what happens in a Nonlinear Context ? *British Journal of Mathematical and Statistical Psychology, 45,* 177-196.

Gregson, R. A. M. (1993a) Learning in the context of nonlinear psychophysics: the Gamma Zak Embedding. *British Journal of Mathematical and Statistical Psychology, 46,* 31-48.

Gregson, R. A. M. (1993b) Minimum Boltzman Entropy and the Identification of n-Dimensional Psychophysics. *Proceedings of the Second Australian Cognitive Science Conference,* Melbourne. pp. 82-84.

Gregson, R. A. M. (1993c) The form of isosimilarity contours in nonlinear psychophysics. *Proceedings of the I S P Conference, Palma, Mallorca.* pp. 51-56.

Gregson, R. A. M. and Britton, L. A. (1990) The size-weight illusion in 2-D nonlinear psychophysics. *Perception and Psychophysics, 48,* 343 - 356.

Gregson, R. A. M. (1994) Similarities derived from 3-d Nonlinear Psychophysics. Variance Distributions. *Psychometrika, 59,* 97-110.

Gregson, R. A. M. and Britton, L. A. (1991) The Size-Weight Illusion in 2-d Nonlinear Psychophysics. *Perception and Psychophysics, 48,* 343-356.

Gregson, R. A. M. and Campbell, E. A. (1991) Two-dimensional Networks in Nonlinear Psychophysical Dynamics. *Proceedings of the 2nd Australian Neural Networks Conference.* Sydney, NSW: Electrical Engineering, Sydney University. pp 215 -222.

Gregson, R. A. M. and Harvey, J. P. (1992) Similarities of Auditory Low-Dimensional Chaotic Sequences to Random Noise. *Perception and Psychophysics, 51,* 267-278.

Gregson, R. A. M. and Wiggins, M. W. (1990) Nonlinear Psychophysics of one Output Component in a Ternary Odour Mixture; Contextual and Sequential Effects. *In* Müller, F. (Ed.) *Fechner Day '90.* Würzburg: Universität. pp 77-82.

Grenander, U. (1993) *General Pattern Theory.* Oxford: Clarendon Press.

Grossberg, S, and Mingolla, E. (1985) Neural dynamics of form perception: Boundary completion, illusory figutes, and neon color spreading. *Psychological review, 92,* 173-211.

Grüsser, O.-J. (1993) The discovery of the psychophysical power law by Tobias Mayer in 1754 and the psychophysical hyperbolic law by Ewald Hering in 1874. *Behavioral and Brain Sciences, 16:1,* 142-144.

Guckenheimer, J. and Holmes, P. (1983) *Nonlinear Oscillations, Dynamical Systems, and Bifurcations of Vector Fields.* New York: Springer-Verlag.

Gull, S. F. and Skilling, J. (1984) The Maximum Entropy Method. *In* Gumbel, E. J. (1958) *Statistics of Extremes.* New York: Columbia University Press.

Gumowski, I. and Mira, C. (1980). *Dynamique Chaotique.* Toulouse: Cepadues Editions.

Harary, F. (1972) *Graph Theory.* New York: Addison Wesley.

Heath, R. A. (1981) A Tandem Random Walk Model for Psychological Discrimination. *British Journal of Mathematical and Statistical Psychology, 34,* 76-92.

Heath, R. A. (1982) A Model for Signal Detection based on an Adaptive Filter. *Biological Cybernetics, 45,* 95-100.

Heinemann, E. G. (1955) Simultaneous brightness induction as a function of inducing- and test- field luminances. *Journal of Experimental Psychology, 50,* 89-96.

Heinemann, E. G. (1972) Simultaneous Brightness Induction. *In* Jameson, D. and Hurvich, L. M. *Visual Psychophysics. (Handbook of Sensory Physiology, Vol. VII/4.)* Berlin: Springer-Verlag. pp 146-169.

Hellström, Å. (1985) The Time-Order and Its Relatives: Mirrors of Cognitive Processes in Comparing. *Psychological Bulletin, 97,* 35-61.

Helson, H. (1933) The fundamental propositions of Gestalt psychology. *Psychological Review, 40,* 13-32.

Hicks, R. E., Miller, G. W., Gaes, G. and Bierman, K. (1977) Concurrent processing demands and the experience of time-in-passing. *American Journal of Psychology, 90,* 431-446.

Hicks, R. E. (1992) Prospective and retrospective judgments of time: A neurobehavioral analysis. *In* Macar, F., Pouthas, V. and Friedman, W. J. (1992) *Time, Action and Cognition.* Dordrecht: Kluwer, pp. 97-108.

Hills, S. (1987) Reference priors and identifiability problems in nonlinear models. *The Statistician, 36,* 235-240.

Hochberg, J. (1984) Form Perception: Experience and Explanations. *In* Dodwell, P. C. and Caelli, T. *Figural Synthesis.* Hillsdale, NJ.: L. Erlbaum Associates, chapter 1.

Hopf, E. (1954) The general temporally discrete Markoff process. *Journal of Rational Mechanics and Analysis, 3,* 13-45.

Helson, H. (1964) *Adaptation Level Theory.* New York: Harper and Row.

Herbart, J. F. (1812) Psychologische Untersuchung ueber die Staerke einer gegebenen Vorstellung, als Function ihrer Dauer betrachtet. *In* Hartenstein, G. (1889) (Ed.) *Johann Friderich Herbart's Schriften zur Psychologie, Dritter Theil.* Hamburg und Leipzig: Leopold Voss.

Heydt, R. van der (1987) Approaches to visual cortical function. *Revues in Physiology, Biochemistry and Pharmacology, 108,* 69-105.

Hopf, E. (1954) The general temporally discrete Markoff process. *Journal of Rational Mechanics and Analysis, 3,* 13-45.

Hornik, K., Stinchcombe, M. and White, H. (1989) Multilayer Feedforward Networks are Universal Approximators. *Neural Networks, 2,* 359-366.

Hornik, K., Stinchcombe, M. and White, H. (1990) Universal Approximation of an Unknown Mapping and Its Derivatives Using Multilayer Feedforward Networks. *Neural Networks, 3,* 551-560.

Hubel, D. H. (1963) The visual cortex of the brain. *Scientific American, 209,* 54-62.

Hurvich, L. M. and Jameson, D. (1966) *The Perception of Brightness and Darkness.* Boston: Allyn and Bacon.

James, W. (1890) *The Principles of Psychology, Vol I.* New York: Henry Holt.

Jamieson, D. G. and Petrusic, W. M. (1975) Presentation order effects in duration discrimination. *Perception and Psychophysics, 17,* 197-202.

Jaynes, (1957) Information Theory and Statistical Mechanics, I.*Physical Review, 106,* 620-630.

Johansson, R. (1976) Skin Mechanoreceptors in the Human Hand: Receptive Field Characteristics. *In* Zotterman, Y. *Sensory Function of the Skin in Primates with special reference to Man.* Oxford: Pergamon Press. pp 159-170.

Johansson, R. and Vallbo, Å. B. (1976) Skin Mechanoreceptors in the Human Hand: An investigation of some Population Properties. *In* Zotterman, Y. *Sensory Function of the Skin in Primates with special reference to Man.* Oxford: Pergamon Press. pp 171-184.

Judd, C. H. (1899) A study of geometrical illusions. *Psychological Review, 6,* 241-261.

Judd, C. H. (1902) Practice and its effect on the perception of illusions. *Psychological Review, 9,* 27-59.

Jumarie, G. (1984) A new approach to the transinformation of random experiments involving fuzzy information. *In* Trappl, R. (Ed.) *Cybernetics and Systems Research 2.* Amsterdam: North Holland.

Justice, J. H. (Ed.) (1986) *Maximum Entropy and Bayesian Methods in Applied Statistics.* Cambridge: Cambridge University Press.

Kabe, D. G. (1966) Complex analogues of some classical noncentral multivariate distributions. *Australian Journal of Statistics, 8,* 99-103.

Kadlec, H. and Townsend, J. T. (1992) Signal Detection Analysis of Dimensional Interactions. *In* Ashby, F. G. (Ed.) *Multidimensional Models of Perception and Cognition.* Hillsdale, N. J.: L. Erlbaum Associates. pp. 181-227.

Kailath, T. (1981) *Lectures on Wiener and Kalman filtering. (International Centre for Mechanical Sciences: CISM Course of Lectures No 40.)* Wien: Springer-Verlag.

Kanizsa, G. (1979) *Organization in Vision: Essays on Gestalt Perception.* New York: Praeger.

Kapitaniak, T. (1990) *Chaos in Systems with Noise, 2nd Edn.* Singapore: World Scientific.

Keyfitz, B. L. and Shearer, M. (1990) *Nonlinear Evolution Equations that Change Type.* New York: Springer-Verlag.

Killeen, P. R. (1992) Counting the minutes. *In* Macar, F., Pouthas, V. and Friedman, W. J. (1992) *Time, Action and Cognition.* Dordrecht: Kluwer, pp. 203-214.

Klien, H. (1919) Beitrag zur Psychopathologie und Psychologie des Zeitsinns. *Zeitschrift für Pathopsychologie, 3* 307-362.

Köhler, W. (1920)*Physiche Gestalten in Ruhe und in stationären Zustand.* Braunschweig: Vieweg.

Köhler, W. (1923) Zur Theorie des Sukzessivvergleichs und der Zeitfehler. *Psychologisches Forschung, 4,* 115-140.

Koster, W. G. and van Schuur, R. (1973) The Influence of the Intensity of Tone Bursts on the Psychological Refractory Period. *In* Kornblum, S. (Ed.) *Attention and Performance, IV.* New York: Academic Press. pp. 55-69.

Krumhansl, C. L. (1978) Concerning the Applicability of Geometric Models to Similarity Data; the Interrelationship between Similarity and Spatial Density . *Psychological Review, 85,* 445-463.

Kullback, S. (1978) *Information Theory and Statistics.* Gloucester, Mass: Peter Smith.

Kullback, S. and Leibler, R. A. (1951) On information and sufficiency. *Annals of Mathematical Statistics, 22,* 79-86.

Langton, C. (1986) Studying artificial life with cellular automata. *Physica, 22D,* 120-149.

Libet, B. (1993) The neural time factor in conscious and unconscious events. *In* Bock, G. R. and Marsh, J. (Eds.) *Experimental and Theoretical Studies of Consciousness. Ciba Foundation Symposium 174.* Chichester: Wiley. pp. 123-146.

Lingoes, J. C. and Roskam, E. E. (1973) A Mathematical and Empirical Analysis of two Multidimensional Scaling Algorithms. *Psychometrika Monograph Supplement 38, (4,2),* 1-81.

Lipps, T. (1883) *Grundtatsuchen des Seelenlebens.* Bonn: Max Cohen und Sohn.

Loftus, G. R., Duncan, J. and Gehrig, P. (1992) On the Time Course of Perceptual Information That Results From a Brief Visual Presentation. *Journal of Experimental Psychology: Human Perception and Performance, 18,* 530-549.

London, I. D. (1954) Research in sensory interaction in the Soviet Union. *Psychological Bulletin, 51,* 531-568.

Luchins, A. S. and Luchins, E. H. (1965) *Logical Foundations of Mathematics for Behavioral Scientists.* New York : Holt Reinhart.

Macar, F., Pouthas, V. and Friedman, W. J. (1992) *Time, Action and Cognition.* Dordrecht: Kluwer.

Mach, E. (1866) Über den physiologischen Effect räumlich vertheilter Lichtreize (zweite Abhandlungen). *Sitzungsberichte der mathematische- naturwissenschaften Classe der kaiserlichen Akademie der Wissenschaften, 54,* 131-144.

Macphail, E. M. (1993) *The Neuroscience of Animal Intelligence.* New York: Columbia University Press.

Mahar, D. and Gregson, R.A.M. (1994) The application of nonlinear psychophysics to the sensory processing of multidimensional tactile patterns. Paper presented at Third International Conference on Tactile Aids, Hearing Aids and Cochlear Implants. Miami, FL.

Mahar, D. P. and Mackenzie, B. D. (1992, in press) Masking, Information Integration and Tactile Pattern Perception: A Comparison of the Isolation and Integration Hypotheses. *Perception,*

Marmarelis, P. Z. and Marmarelis, V. Z. 91978) *Analysis of Physiological Systems.* New York: Plenum Press.

McGee, V. E. (1971) *Principles of Statistics: Traditional and Bayesian.* New York: Appleton-Century-Crofts.

McNaughton, B. L. and Morris, R. G. M. (1987) Hipppocampal synaptic enhancement and information storage within a distributed memory system. *Trends in Neuroscience, 10,* 408-415.

Melara, R. D. (1992) The Concept of Perceptual Similarity: From Psychophysics to Cognitive Psychology. *In* Algom, D. (Ed.) *Psychophysical Approaches to Cognition.* Amsterdam: Elsevier.

Melara, R. D. and Day, D. J. A. (1992) Primacy of dimensions in vibrotactile perception: An evaluation of early holistic models. *Perception and Psychophysics, 52,* 1-17.

Melara, R. D. and Marks, L.E.(1990) Perceptual primacy of dimensions: Support for a model of dimensional interaction. *Journal of Experimental Psychology: Human Perception and Performance, 16,* 398-414.

Metzger, W. Vukovich-Voth, O., and Koch, I. (1970) Über optisch-haptische Masstäuschungen an dreidimensionalen Gegenständen. *Psychologische Beiträge, 18,* 329-366.

Mey, H. (1972) *Field-theory: a study of its application in the social sciences.* London: Routledge and Kegan Paul.

Mielniczuk, J. and Tyrcha, J. (1993) Consistency of Multilayer Perceptron Regression Estimators. *Neural Networks, 6,* 1019-1022.

Miller, K. S. (1968) Moments of complex gaussian processes. *IEEE Proceedings (Letters), 56,* 83-84.

Miller, K. S. (1974) *Complex Stochastic Processes.* Reading, Mass.: Addison-Wesley.

Miller, K. S. and Rochwarger, M. M. (1970) Estimation of spectral moments of time series. *Biometrika, 57,* 513-517.

Mohammed-Djafari, A. and Demoment, G. (1989) Maximum Entropy and Bayesian Approach in Tomographic Image Reconstruction and Restoration. *In* Skilling, J. *Maximum Entropy and Bayesian Methods.* Dordrecht: Kluwer. pp 195-201.

Moreno, E. and Cano, J. A. (1991) Robust Bayesian Analysis with ϵ-contaminations Partially Known. *Journal of the Royal Statistical Society, Series B, 53,* 143-155.

Mori, T. (1981) Reduction of the Poggendorff effect by the motion of oblique lines. *Perception and Psychophysics, 29,* 15-20.

Moulden, B. and Kingdom, F. (1989) White's Effect: A Dual Mechanism. *Vision Research, 29,* 1245-1259.

Murdoch, D. C. (1957) *Linear Algebra for Undergraduates.* London: Chapman and Hall.

Murray, D. (1993) A Perspective for Viewing the History of Psychophysics. *Behavioral and Brain Sciences, 16,* 125-155.

Reading, Mass.: Addison-Wesley.

Nakajima, Y., Nishimura, S. and Teranishi, R.(1988) Ratio judgments of empty durations with numeric scales. *Perception, 17,* 93-118.

Needham, J. G. (1934) The time-error in comparison judgments. *Psychological Bulletin, 31,* 229-243.

Needham, J. G. (1935) The effect of the time interval upon the time error at different intensive levels. *Journal of Experimental Psychology, 18,* 539-543.

Nosofsky, R.M. (1985) Overall Similarity, and the Identification of Separable-dimension Stimuli: A Choice Model Analysis. *Perception and Psychophysics, 38,* 415-432.

Nosofsky, R.M. (1986) Attention, Similarity, and the Identification -categorization relationship. *Journal of Experimental Psychology: General, 115,* 39-57.

Nosofsky, R. M. (1992) Similarity Scaling and Cognitive Process Models. *Annual Review of Psychology, 43,* 25-54.

Nosofksy, R. M. and Smith, A. (1992) Similarity, Identification and Categorization. *Journal of Experimental Psychology: General, 121,* 237-245.

Nummelin, E. (1984) *General irreducible Markov chains and non-negative operators.* Cambridge: Cambridge Universtiy Press.

Nunez, P. L. (Ed.) (1995) *Neocortical Dynamics and Human EEG Rhythms.* Oxford: Oxford University Press.

Ott, E. (1993) *Chaos in Dynamical Systems.* Cambridge: Cambridge University Press.

Over, R. (1968) Explanations of geometric illusions. *Psychological Bulletin, 70,* 545-562.

Peitgren, H.-O., Jürgens, H. and Saupe, D. (1992) *Chaos and Fractals.* New York: Springer-Verlag.

Petermann, B. (1932) *The Gestalt Theory and the Problem of Configurations.* London: Routledge and Kegan Paul.

Petry, S. and Meyer, G. E. (Eds.) (1987) *The Perception of Illusory Contours.* New York: Springer-Verlag.

Piaget, J. (1961) *Les Mechanismes Perceptifs.* (trans. Seagrim, G. N. as *The Mechanisms of Perception.* London: Kegan Paul, 1969).

Poulton, E. C. (1989) *Bias in Quantifying Judgments.* Hove: Lawrence Erlbaum Associates.

Press, W.H., Flannery, B. P., Teukowsky, S. A., and Vetterling, W. T. (1986). *Numerical recipes.* Cambridge: Cambridge University Press.

Pressey, A. W. and Pressey, C. A. (1992) Attentive fields are related to focal and contextual features; A study of Müller-Lyer distortions. *Perception and Psychophysics, 51,* 423-436.

Price, I. R. and Gregson, R. A. M. (1988) Nonlinear Dynamics in a Complex Cubic One-dimensional Model for Sensory Psychophysics. *Acta Applicandae Mathematicae, 11,* 1 17.

Pritchard, R. M., Heron, W. and Hebb, D. O. (1960) Visual perception approached by the method of stabilized images. *Canadian Journal of Psy-*

chology, 14, 67-77.

Pylyshyn, Z. W. (1980) Computation and cognition: Issues in the foundations of cognitive science. *The Behavioral and Brain Sciences, 3,* 111-169.

Radeau, M. (1992) Cognitive Impenetrability in Auditory-Visual Interaction. *In* Alegria, J., Holender, D., Junça de Morais, J. and Radeau, M. (Eds.) *Analytic Approaches to Human Cognition.* Amsterdam: North Holland. pp. 41-55.

Ratliff, F. (1965) *Mach Bands: quantitative studies on neural networks in the retina.* San Francisco: Holden Day.

Rausch, E. (1966) Problem der Metrik (Geometrisch-optische Täuschungen. *In :* Metzger, W. (Ed.) *Handbuch der Psychologie, Band I.* Göttingen: Hogrefe.

Ratliff, F. (1965) *Mach Bands: quantitative studies on neural networks in the retina.* San Francisco: Holden Day.

Reneke, J. A., Fennell, R. E. and Minton, R. B. (1987) *Structured Hereditary Systems.* New York: Marcel Dekker.

Ripley, B. D. (1994) Neural Networks and Related Methods for Classification. *Journal of the Royal Statistical Society, Series B, 56,* 409-456.

Roberts, J. A. (Ed.) (1989) *Indirect Imaging.* Cambridge: Cambridge University Press.

Robinson, J. O. (1972) *The Psychology of Visual Illusion.* London: Hutchinson.

Rolls, E. T. (1989) Parallel distributed processing in the brain; implications of the functional architecture of neuronal networks in the hippocampus. *In:* Morris, R. G. M. (Ed.) *Parallel Distributed Processing: Implications for Psychology and Neurobiology.* Oxford: Clarendon Press. pp. 286-388.

Rosenblith, W. A. *Sensory Communication.* New York: Wiley.

Rubin, (1950) Visual Figures apparently incompatible with Geometry. *Acta Psychologia, 7,* 365-387.

Saitou, S. (1988) *Theory of Reproducing Kernels and its Applications.* Harlow, Essex: Longman.

Scharf, B. (Ed.) (1975) *Experimental Sensory Psychology.* Glenview, Ill.: Scott, Foresman.

Schetzen, M. (1980) *The Volterra and Wiener Theories of Nonlinear Systems.* New York: Wiley.

Schmitt, S. A. (1969) *Measuring Uncertainty: An Elementary Introduction to Bayesian Statistics.* Reading, Mass.: Addison-Wesley.

Silva, J. A. L. and Hallam, T. G. (1992) Compensation and Stability in Nonlinear Matrix Models. *Mathematical Biosciences, 110,* 67-101.

Shepard, R. N. (1974) Representation of Structure in Similarity Data: Problems and Prospects. *Psychometrika, 39,* 373-421.

Shepard, R. N. (1991) Integrability versus Separability of Stimulus Dimensions: From an Early Convergence of Evidence to a Proposed Theoretical Basis. *In* Lockhead, G. R. and Pomerantz, J. R. (Eds.) *The Perception of Structure.* Washington, DC: American Psychological Association. pp. 53-72.

Shore, J. E. and Johnson, R. W. (1980) *IEEE Transactions on Information Theory, IT-26,* 26-37.

Skara, C. A., and Freeman, W. J. (1988) How brains make chaos in order to make sense of the world. *Behavioral and Brain Sciences, 10,* 161-196.

Sierpinski, W. (1956) *General Topology.* Toronto, University of Toronto Press.

Skilling, J. (1984) The maximum entropy method. *Nature, 309,* 748-749.

Skilling, J. (Ed.) (1989a) *Maximum Entropy and Bayesian Methods.* Boston: Kluwer.

Skilling, J. (1989b) Classic Maximum Entropy. *In* Skilling, J. (Ed.) *Maximum Entropy and Bayesian Methods.* Boston: Kluwer. pp. 45-52.

Smith, C. R. and Erickson, G. J. (Eds.) (1987) *Maximum Entropy and Bayesian Spectral Analysis and Estimation Problems.* Dordrecht: Reidel.

Srivastava, H. M. and Buschman, R. G. (1977) *Convolution Integral Equations with Special Function Kernels.* New York: Wiley.

Stevens, S. S. (1939) Tactile Vibration: Dynamics of Sensory Intensity. *Journal of Experimental Psychology, 57,* 210-218.

Thompson, J. M. T. and Stewart, H. B. (1986) *Nonlinear Dynamics and Chaos.* Chichester: Wiley.

Tikochinsky, Y., Tishby, N. Z. and Levine, R. D. (1984) Consistent Inference of Probabilities for Reproducible Experiments. *Physical Review Letters, 52,* 1357-1360.

Torgerson, W. S. (1962) *Theory and Methods of Scaling.* New York: Wiley.

Townsend, J. T. (1972) Some results concerning the identification of parallel and series processes. *British Journal of Mathematical and Statistical Psychology, 25,* 169-199.

Townsend, J. T. and Ashby, E. G. (1983) *Stochastic Modeling of Elementary Psychological Processes.* Cambridge: Cambridge University Press.

Townsend, J. T., Hu, G. G. and Evans, R. J. (1984) Modelling Feature Perception in Brief Displays with Evidence for Positive Interdependence. *Perception and Psychophysics, 36,* 35-49.

Treisman, M., Cook, N., Naish, P. L. N. and MacCrone, J. K. (1994) The Internal Clock: Electroencephalograpic Evidence for Oscillatory Processes

Underlying Time Perception. *Quarterly Journal of Experimental Psychology, 47A,* 241-289.

Tribus,M. (1969) *Rational Descriptions, Decisions, and Designs.* New York: Pergamon.

Tu, P. N. V. (1994) *Dynamical Systems, 2nd Edn.* Berlin: Springer-Verlag.

Tversky, A. (1977) Features of Similarity. *Psychological Review, 84,* 327 - 352.

Tversky, A. and Gati, D. (1982) Similarity, Separability, and the Triangle Inequality. *Psychological Review, 89,* 123-154.

Urmson, J. O. (1992) *Simplicius On Aristotle Physics 4.1-5, 10-14.* London: Duckworth.

Vallbo, Å. B. and Johansson, R. (1976) Skin Mechanoreceptors in the Human Hand: Neural and Psychophysical Thresholds. *In* Zotterman, Y. *Sensory Function of the Skin in Primates with special reference to Man.* Oxford: Pergamon Press. pp 185-200.

Vardi, Y. and Lee, D. (1993) From Image Deblurring to Optimal Investments: Maximum Likelihood Solutions for Positive Linear Inverse Problems. *Journal of the Royal Statistical Society, Series B, 55,* 569-612.

Vickers, D. (1979) *Decision Processes in Visual Perception.* New York: Academic Press.

Vierordt, K. (1868) *Der Zeitsinn nach Versuchen.* Tübingen: Laupp.

von Fieandt, K. and Moustgaard, I. K. (1977) *The Perceptual World.* London: Academic Press.

von Neumann, J. (1954) *The Computer and the Brain.* Yale Univ. Press.

Waldrop, M. M. (1992) *Complexity.* London: Penguin Books.

Wall, P. D. (1961) Two Transmission Systems for Skin Sensations. *In* Rosenblith, W. A. *Sensory Communication.* New York: Wiley. pp 475-496.

Wall, P. D. and Cronly-Dillon, J. (1960) Pain, Itch and Vibration. *A.M.A. Archives of Neurology, 2,* 365-375.

Walley, P. (1991) *Statistical Reasoning with Imprecise Probabilities.* London: Chapman and Hall.

Walsh, J. E. (1965) *Handbook of Nonparametric Statistics, Vol. II* Princeton: Van Nostrand.

Ward, L. M. and Coren, S. (1976) The effect of optically-induced blur on the magnitude of the Müller-Lyer illusion. *Bulletin of the Psychonomic Society, 7,* 483-484.

Watson, A. B. (1986) Temporal Sensitivity. *In* Boff, K. R., Kaufman, L. and Thomas, J. P. (Eds) *Handbook of Perception and Human Performance, Vol. I: Sensory Processes and Perception.* New York: Wiley. Chapter 6.

Wearden, J. H. and Ferrara, A. (1993) Subjective Shortening in Humans' Memory for Stimulus Duration. *Quarterly Journal of Experimental Psychology, 46B,* 163-186.

Weber, E. H. (1848) *De pulsu, resorptione, auditu et tactu.* Leipzig: Koehler.

Weintraub, D. J. and Virsu, V. (1971) The misperception of angles: estimating the vertex of converging line segments. *Perception and Psychophysics, 9,* 5-8.

Welch, R. B. and Warren, D. H. (1980) Immediate perceptual response to intersensory discrepancy. *Psychological Bulletin, 88,* 638-667.

Welch, R. B. and Warren, D. H. (1986) Intersensory Interaction. *In* Boff, K. R., Kaufman, L. and Thomas, J. P. (Eds.) *Handbook of Perception and Human Performance, Vol. I: Sensory Processes and Perception.* New York: Wiley. Chapter 25.

Wertheimer, M. (1923) Untersuchungen zur Lehre von der Gestalt, II. *Psychologische Forschung, 4,* 301-350.

White, H. (1990) Connectionist Nonparametric Regression: Multilayer Feedforward Networks Can Learn Arbitrary Mappings. *Neural Networks, 3,* 539-549.

Wickelgren, W. A. (1972) Trace-resistance and the decay of long-term memory. *Journal of Mathematical Psychology, 9,* 418-455.

Wiggins, S. (1992) *Chaotic Transport in Dynamical Systems.* Berlin: Springer-Verlag.

Wiles, J., Lippiatt, G. and Broerse, J. (1990) An Explanation of Misalignment in the Poggendorf Illusion in Terms of Simple Line Completion Processes. *In* Jabry, M. (Ed.) *Proceedings of the First Australian Conference on Neural Networks,* Electrical Engineering, Sydney University, NSW, pp 120-122.

Wolf, A., Swift, J. B., Swinney, H. L. and Vastano, J. A. (1985) Determining Lyapunov Exponents from a Time Series. *Physica, 16D,* 285-317.

Wolfram, S. (1984) Twenty Problems in the Theory of Cellular Automata. *Proceedings of the Nobel Symposium 59: The Physics of Chaos and Related Problems.* Graftvallen, Sweden.

Wolfram, S. (1986) *Theory and Applications of Cellular Automata.* Singapore: World Scientific.

Wooding, R. A. (1956). The multivariate distribution of complex normal variables. *Biometrika, 43,* 212-215.

Woodrow, H. (1951) *Time perception.* In: Stevens, S. S. (Ed.) *Handbook of Experimental Psychology.* New York: Wiley.

Young, F. W. (1970) Nonmetric Multidimensional Scaling: Recovery of Metric Information. *Psychometrika, 35,* 455-473.

Zakay, D. (1990) The Evasive Art of Subjective Time Measurement: Some Methodological Dilemmas. *In* Block, R. A. (1990) *Cognitive Models of Psychological Time.* Hillsdale, NJ: Lawrence Erlbaum Associates, pp. 59-84.

Zakay, D., Nitzam, D. and Glicksohn, J. (1983) The influence of task difficulty and external tempo on subjective time estimation. *Perception and Psychophysics, 34,* 451-456.

Zelkind, I., and Sprug, J. (1974) *Time Research: 1172 Studies.* Metuchen, NJ: Scarecrow.

Zigler, M. J. (1935) The experimental relation of the two-point limen to the error of localisation. *Journal of General Psychology, 13,* 316-332.

Zwicker, E. and Fastl, H. (1990) *Psychoacoustics; Facts and Models.* New York: Springer-Verlag.

Subject Index

V:

W:

www.ingramcontent.com/pod-product-compliance
Ingram Content Group UK Ltd.
Pitfield, Milton Keynes, MK11 3LW, UK
UKHW020358150926
8624IPUK00004B/230

9 789810 223274